ASKING QUESTIONS IN BIOLOGY

ASKING QUESTIONS IN BIOLOGY

Design, Analysis and Presentation in Practical Work

Chris Barnard, Francis Gilbert
& Peter McGregor

Department of Life Science
University of Nottingham

 LONGMAN

Addison Wesley Longman Limited,
Edinburgh Gate
Harlow
Essex CM20 2JE
United Kingdom
and Associated Companies throughout the world.

First edition 1993
Reprinted 1994, 1995, 1996 (twice)

British Library Cataloguing in Publication Data

A catalogue record for this book is available from the British Library

ISBN 0-582-08854-2

Library of Congress Cataloging-in-Publication Data

Barnard, C. J. (Christopher John)
   Asking questions in biology : design, analysis, and presentation
in practical work / Chris Barnard, Francis Gilbert & Peter McGregor.
— 1st ed.
     p.   cm.
   Includes bibliographical references and index.
   1. Science—Methodology. 2. Biology—Methodology. I. Gilbert
Francis S. (Francis Sylvest), 1956– . II. McGregor, Peter K.
(Peter Kenneth)
III. Title.
Q175.B218   1993
574'.028—dc20

Set by 6 in 10/12pt Plantin

Produced through Longman Malaysia, TCP

CONTENTS

PREFACE

Science is a process of asking questions, in most cases precise, quantitative questions that allow distinctions to be drawn between alternative explanations of events. Asking the right questions in the right way is a fundamental skill in scientific enquiry, yet in itself it receives surprisingly little explicit attention in scientific training. Students being trained in scientific subjects, for instance in sixth forms, colleges and universities, learn the factual science and some of the tools of enquiry such as laboratory techniques, mathematics, statistics and computing, but they are taught little about the process of question-asking itself.

This book has its origins in a first-year practical course we have run at Nottingham over the past few years. The aim of the course is to introduce students in the life sciences to the skills of observation and enquiry, but focusing on the process of enquiry – how to formulate hypotheses and predictions from raw information, how to design critical observations and experiments and how to choose appropriate analysis – rather than on laboratory, field and analytical techniques *per se*. Statistics and the use of computer packages are introduced simply as aids at the appropriate stage of enquiry. We thus emphasize their role and the information they provide rather than the theory behind them and the mechanics of their use. Provided students know which test to use and how to interpret its result, statistics can be regarded as a useful 'black box'.

The book takes this approach by looking at the process of enquiry during its various stages, starting with unstructured observations and working through to the production of a complete written report. In each section, different skills are emphasized and a series of main examples runs through the book to illustrate their application at each stage. We have deliberately used behaviour as a vehicle for the main examples because of its flexibility and scope for open-ended investigation and because it is cheap and easy to lay on as practical material for large numbers of students. However, as should be readily apparent from the range of other examples used, the vehicle itself is immaterial to the aims of the book, which is concerned with asking questions across the field of biology.

The book begins with a look at scientific question-asking in general. How do we arrive at the right questions to ask? What do we have to know before we can ask sensible questions? How should questions be formulated to be answered most usefully? The section addresses these points by looking at the development of testable hypotheses and predictions and the sources from which they might arise.

The next section looks at how hypotheses and predictions can be derived from unstructured observational notes. Exploratory analysis is an important first step in deriving hypotheses from raw data and the section introduces plots and summary statistics as useful ways of identifying interesting patterns on which to base hypotheses. The section concludes by pointing out that although hypotheses and their predictions are naturally specific to the investigation in hand, testable predictions in general fall into two distinct groups: those dealing with some kind of **difference** between groups of data and those dealing with a **trend** in the quantitative relationship between groups of data.

The distinction between difference and trend predictions is developed further in the third section which discusses the use of confirmatory analyses. The concept of statistical significance is introduced as an arbitrary but generally agreed yardstick as to whether observed differences or trends are interesting and a number of basic but broadly applicable significance tests are explained. Throughout, however, the emphasis is on the **use** of such tests as tools of enquiry rather than on the statistical theory underlying them. Having introduced significance tests and some potential pitfalls in their use, the book uses the main worked examples to show how some of their predictions can be tested and hypotheses refined in the light of testing.

In the final section, the book considers the presentation of information. Once hypotheses have been tested, how should the outcome be conveyed for greatest effect? The section discusses the use of tables, figures and other modes of presentation and shows how a written report should be structured. The points made in the section are then illustrated in a complete written report based on one of the main worked examples.

At the end of the book is a number of appendices. These provide some self-test questions and answers based on the material in the book, some worked examples of significance tests and some statistical tables for use in significance testing.

We said that the book had its inception in our introductory practical course. However, the practical course itself was developed in response to an increasingly voiced need on the part of students to be taught how to ask and answer questions. Both the practical course and the book have benefited immensely from a constant and pleasurable interaction with our undergraduates over the years. Their enquiries and insights continue to hone the way we teach and have been the guiding force behind all the discussions in the book. Without them it simply wouldn't have been written.

CJB, FSG, PKM

1

DOING SCIENCE

You're out for a walk one autumn afternoon when you notice a squirrel picking up acorns under some trees. Several things strike you about the squirrel's behaviour. For one thing it doesn't seem to pick up all the acorns it comes across; a sizeable proportion is ignored. Of those it does pick up, only some are eaten. Others are carried up into a tree where the squirrel disappears from view for a few minutes before returning to the supply for more. Something else strikes you: the squirrel doesn't carry its acorns up the nearest tree but instead runs to one several metres away. You begin to wonder why the squirrel behaves in this way. Several possibilities occur to you. Although the acorns on the ground all look very similar to you, you speculate that some might contain more food than others, or perhaps they are easier to crack. By selecting these, the squirrel might obtain food more quickly than by taking indiscriminately any acorn it encountered. Similarly, the fact that it appears to carry acorns into a particular tree suggests this tree might provide a more secure site for storing them.

While all these might be purely casual reflections, they are revealing of the way we analyse and interpret events around us. The speculations about the squirrel's behaviour may seem clutched out of the air on a whim but they are in fact structured around some clearly identifiable assumptions, for instance that achieving a high rate of food intake matters in some way to the squirrel and influences its preferences, and that using the most secure storage site is more important to it than using the most convenient site. If you wanted to pursue your curiosity further, these assumptions would be critical to the questions you asked and the investigations you undertook. If all this sounds very familiar to you as a science student it should because, whether you intended it or not, your speculations are essentially scientific. Science is simply formalized speculation backed up (or otherwise) by equally formalized observation and experimentation. In its broadest sense most of us 'do science' all the time.

SCIENCE AS ASKING QUESTIONS

Science is often regarded by those outside it as a open-ended quest for objective understanding of the universe and all that is in it. But this is so only in a rather trivial sense. The issue of objectivity is a thorny one and happily well beyond the scope of this book. Nevertheless, the very real constraints that limit human objectivity mean that use of the term must at least be hedged about with serious qualifications. The issue of open-endedness is really the one that concerns us here. Science is open-ended only in that its directions of enquiry are in principle limitless. Along each path of enquiry, however, it is far from open-ended. Each step on the way is, or should be, the result of refined question-asking, a narrowing down of questions and methods of answering them to provide the clearest possible distinction between alternative explanations for the phenomenon in hand. This is a skill, or series of skills really, that has to be acquired, and acquiring it is one of the chief objectives of any scientific training.

While few scientists would disagree with this, identifying the different skills and understanding how training techniques develop them are a lot less straightforward. With increasing pressure on science courses in universities and colleges to teach more material to more people and to draw on an expanding and increasingly sophisticated body of knowledge, it is more important than ever to understand how to marshal information and direct enquiry. This book is the result of our experiences in teaching investigative skills to university undergraduates in the life sciences. It deals with all aspects of scientific investigation, from thinking up ideas and making initial exploratory observations, through developing and testing hypotheses to interpreting results and preparing written reports. It is not an introduction to data-handling techniques or statistics, although it includes a substantial element of both; it simply introduces these as tools to aid investigation. The theory and mechanics of statistical analysis can be dealt with more appropriately elsewhere.

The principles covered in the book are extraordinarily simple, yet paradoxically students find them very difficult to put into practice when taught in a piecemeal way across a number of different courses. The book has evolved out of our attempts to get over this problem by using open-ended, self-driven practical exercises in which the stages of enquiry develop logically through the desire by students to satisfy their own curiosity. However, the skills it emphasizes are just as appropriate to more limited set-piece practicals. Perhaps a distinction – admittedly over-generalized – that could be made here, and which to some extent underpins our preference for a self-driven approach, is that with many set-piece practicals it is obvious **what** one is supposed to do but often not **why** one is supposed to do it. Almost the opposite is true of the self-driven approach; here it is clear why any investigation needs to be undertaken but usually less clear what should be done to see it through successfully. In our experience, developing the 'what' in the context of a clear 'why' is considerably more instructive than attempting to reconstruct the 'why' from the 'what' or, worse, ignoring it altogether.

BASIC CONSIDERATIONS

Scientific enquiry is not just a matter of asking questions; it is a matter of asking the **right questions** in the **right way**. This is more demanding than it sounds. For a start, it requires that something is known about the system or material in which an investigator is interested. A study of mating behaviour in guppies, for instance, demands that you can at least tell males from females and recognize courtship and copulation. Similarly, it is difficult to make a constructive assessment of parasitic worm burdens in host organisms if you are ignorant of likely sites of infection and can't tell worm eggs from faecal material.

Of course, there are several ways in which such knowledge can be acquired: e.g. textbooks, specialist journals, asking an expert or simply finding out for yourself through observation and exploration. Whichever way is most appropriate, however, a certain amount of background preparation is usually essential, even for the simplest investigations. In practical classes, some background is usually provided for you in the form of handouts or accompanying lectures, but the very variability of biological material means that generalized and often highly stylized summaries are poor substitutes for hard personal experience. Nevertheless, given the inevitable constraints of time, materials and available expertise, they are usually a necessary second best. There is also a second, more important, reason why there is really no substitute for personal experience: the information you require may not exist or, if it does exist, it may not be correct. Taking received wisdom at face value can be a dangerous business – something even seasoned researchers can continue to discover, the famous geneticist and biostatistician R. A. Fisher among them.

In the early 1960s, Fisher and other leading authorities at the time were greatly impressed by an apparent relationship between duodenal ulcer and certain rhesus and MN blood groups. Much intellectual energy was expended in trying to explain the relationship. A sceptic, however, mentioned the debate to one of his blood-group technicians. The technician, for years at the sharp end of blood-group analysis, resolved the issue on the spot. The relationship was an artefact of blood transfusion! Patients with ulcers had received transfusions because of haemorrhage. As a result, they had temporarily picked up rhesus and MN antigens from their donors. When patients who had not been given transfusions were tested, the relationship disappeared (Clarke, 1990).

Where at all feasible, therefore, testing assumptions yourself and making up your own mind about the facts available to you is a good idea. It is impossible to draw up a definitive list of what it is an investigator needs to know as essential background; biology is too diverse a subject and every investigation is to some extent unique in its factual requirements. Nevertheless, it is useful to indicate the kinds of information that are likely to be important. Some examples might be as follows:

Question

Can the material of interest be studied usefully under laboratory conditions or will unavoidable constraints or manipulations so affect it that any conclusions will have only dubious relevance to its normal state or functions?

For instance, can mating preferences in guppies usefully be studied in a small plastic aquarium, or will the inevitable restriction on movement and the impoverished environment compromise normal courtship activity?
Or, if the growth and developmental fate of certain cells can be monitored only with the aid of a vital dye, will normal function be maintained in the dyed state or will the dye interfere subtly with the processes of interest?

Question

Is the material at the appropriate stage of life history or development for the desired investigation?

There would, for instance, be little point in carrying out vaginal smears on female mice to establish stages of the oestrous cycle if some females were less than 28–32 days of age. Such mice may well not have begun cycling.
Likewise, it would be fruitless to monitor the faeces of infected mice for the eggs of a nematode worm until a sufficient number of days have passed after infection for the worms to have matured.

Will the act of recording from the material affect its performance?

For example, removing a spermatophore (package of sperm
donated by the male) from a recently mated female cricket in
order to assay its sperm content may adversely affect the
female's response to males in the future.
Or, the introduction of an intracellular probe might disrupt the
aspect of cell physiology it was intended to record.

Has the material been prepared properly?

If the problem to be investigated involves a foraging task
(e.g. learning to find cryptic prey), has the subject been trained
to perform in the apparatus and has it been deprived of food for
a short while to make it hungry?
Similarly, if a mouse of strain X is to be infected with a
particular blood parasite so that the course of infection can be
monitored, has the parasite been passaged in the strain long
enough to ensure its establishment and survival in the
experiment?

Does the investigation make demands on the material that it is not
designed to meet?

Testing for the effects of acclimation on some measure of coping
in a new environment might be compromised if conditions in
the new environment are beyond those the organism's
physiology or behaviour have evolved to meet.

Likewise, testing a compound from an animal's environment for carcinogenic properties in order to assess risk might not mean much if the compound is administered in concentrations or via routes that the animal could never experience naturally.

Are assumptions about the material justified?

In an investigation of mating behaviour in dragonflies, we might consider using the length of time a male and female remain coupled as an index of the amount of sperm transferred by the male. Before accepting this, however, it would be wise to conduct some pilot studies to make sure it was true; it might be, for instance, that some of the time spent coupled reflected mate-guarding rather than insemination.

By the same token, assumptions about the relationship between the staining characteristics of cells in histological sections and their physiological properties might need verifying before concluding anything about the distribution of physiological processes within an organ.

The list could go on for a long time, but these examples are basic questions of practicality. They are not very interesting in themselves but they, and others like them, need to be addressed before interesting questions can be asked. Failure to consider them will almost inevitably result in wasted time and materials.

Of course, even at this level, the investigator will usually have the questions ultimately to be addressed – the whole point of the investigation – in mind, and these will naturally influence initial considerations.

THE SKILL OF ASKING QUESTIONS

Testing hypotheses

A famous naturalist once remarked that without a hypothesis a geologist might as well go into a gravel pit and count the stones. He meant, of course, that simply gathering facts for their own sake was likely to be a waste of time. A geologist is unlikely to profit much from knowing the number of stones in a gravel pit. This seems self-evident, but such undirected fact-gathering is a common problem among students in practical and project work. There can't be many science teachers who have not been confronted by a puzzled student with the plea: 'I've collected all these data, now what do I do with them?' The answer, obviously, is that the investigator should know what is to be done with the data before they are collected. As the naturalist well knew, what gives data collection direction is a working **hypothesis**.

The word 'hypothesis' sounds rather formal and, indeed, in some cases hypotheses may be set out in a tightly constructed, formal way. In more general usage, however, its meaning is a good deal looser. Verma and Beard (1981), for example, define it as simply

> a tentative proposition which is subject to verification through subsequent investigation. ... In many cases hypotheses are hunches that the researcher has about the existence of relationships between variables

A hypothesis, then, can be little more than an intuitive feeling about how something works, or how changes in one factor will relate to changes in another, or about any aspect of the material of interest. However vague it may be, though, it is formative in the purpose and design of investigations because these set out to test it. If at the end of the day the results of the investigation are at odds with the hypothesis, the hypothesis may be rejected and a new one put in its place. As we shall see later, hypotheses are never proven, merely rejected if data from tests so dictate, or retained for the time being for possible rejection after yet further tests.

How is a hypothesis tested?

If a hypothesis is correct, certain things will follow. Thus if our hypothesis is that a particular visual display by a male chaffinch is sexual in motivation, we might expect the male to be more likely to perform the display when a female is present. Hypotheses thus generate **predictions**, the testing of which increases or decreases our faith in them. If our male chaffinch turned out to display mainly when other males were around and almost never with females, we might want to think again about our sexual motivation hypothesis. However, we should be wrong to dismiss it solely on these grounds. It could be that such displays are important in defending a good quality breeding territory which eventually will attract a female. The context of the display could thus still be sexual, but in a less direct sense than we had first considered. In this way, hypotheses can produce tiers of more and more refined predictions before they are rejected or tentatively accepted. Making such predictions is a skilled business because each must be phrased so that testing it allows the investigator to discriminate in favour of or against the hypothesis. While it is best to phrase predictions

as predictions (thus: *males will perform more of display* y *in the presence of females*), they sometimes take the form of questions (*do males perform more of display* y *when females are present?*). The danger with the question format, however, is that it can easily become too woolly and vague to provide a rigorous test of the hypothesis (e.g. *do males behave differently when females are present?*). Having to phrase a precise prediction helps counteract the temptation to drift into vagueness.

Hypotheses, too, can be so broad or imprecise that they are difficult to reject. In general the more specific, mutually exclusive hypotheses that can be formulated to account for an observation the better. In our chaffinch example, the first hypothesis was that the display was sexual in motivation. Another might be that it reflected aggressive defence of food. Yet another that it was an anti-predator display. These three hypotheses give rise to very different predictions about the behaviour and it is thus, in principle, easy to distinguish between them. As we have already seen, however, distinguishing between the 'sexual' and 'aggressive' hypotheses may need more careful consideration than we first expect. We shall look at the development of hypotheses and their predictions in more detail later on.

WHERE DO QUESTIONS COME FROM?

As we have already intimated, questions do not spring out of a vacuum. They are triggered by something. They may arise from a number of sources.

Curiosity

Questions arise naturally when thinking about any kind of problem. Simple curiosity about how something works or why one group

of organisms differs in some way from another group can give rise
to useful questions from which testable hypotheses and their
predictions can be derived. There is nothing wrong with 'armchair
theorizing' and 'thought experiments' as long as, where possible,
they are put to the test. Sitting in the bath and wondering about
how migratory birds manage to navigate, for example, could
suggest roles for various environmental cues like the sun, stars and
topographical features. This in turn could lead to hypotheses about
how they are used and predictions about the effects of removing
or altering them. By the time the water was cold, some useful
preliminary experiments might even have been devised.

Casual observation

Instead of dreaming in the bath, you might be watching a tank
full of fish, or sifting through some histological preparations under
a microscope. Various things might strike you. Some fish in the
tank might seem very aggressive, especially towards others of their
own species, but this aggressiveness might occur only around
certain objects in the tank, perhaps an overturned flowerpot or a
clump of weed. Similarly, certain cells in the histological
preparations may show unexpected differences in staining or
structure. Even though these aspects of fish behaviour and cell
appearance were not the original reason for watching the fish or
looking at the slides, they might suggest interesting hypotheses
for testing later. A plausible hypothesis to account for the
behaviour of the fish, for instance, is that the localized aggression
reflects territorial defence. Two predictions might then be: (a) *on
average, territory defenders will be bigger than intruders* (because
bigger fish are more likely to win in disputes and thus obtain a
territory in the first place) and (b) *removing defendable resources
like upturned flowerpots will lead to a reduction in aggressive
interactions.* Similarly, a hypothesis for differences in cell staining
and structure is that they are due to differences in the age and
development of the cells in question. A prediction might then be:
younger tissue will contain a greater proportion of (what are
conjectured to be the) *immature cell types.*

Exploratory observations

It may be that you already have a hypothesis in mind, say that a
particular species of fish will be territorial when placed in an
appropriate aquarium environment. What is needed is to decide
what an appropriate aquarium environment might be so that
suitable predictions can be made to test the hypothesis. Obvious

things to do would be to play around with the size and number of shelters, the position and quality of feeding sites, the number and sex ratio of fish introduced into the tank and so on. While the effects of these and other factors on territorial aggressiveness among the fish might not have been guessed at beforehand, such manipulations are likely to suggest relationships with aggressiveness which can then be used to predict the outcome of further, **independent** investigations. Thus if exploratory results suggested aggressiveness among defending fish was greater when there were ten fish in the tank compared with when there were five, it would be reasonable to predict that aggressiveness would increase as the number of fish increased, *all other things being equal*. An experiment could then be designed in which shelters and feeding sites were kept constant but different numbers of fish, say 2, 4, 6, 8, 10 or 15, were placed in the tank. Measuring the amount of aggression by a defender with each number of fish would provide a test of the prediction.

Previous studies

One of the richest sources of questions is, of course, past and ongoing research. This might be encountered either as published literature or 'live' as research talks at conferences or seminars. A careful reading of most published papers, articles or books will turn up ideas for further work, whether at the level of alternative hypotheses to explain the problem in hand or at the level of further or more discriminating predictions to test the current hypothesis. Indeed, this is the way most of the scientific literature develops. Some papers, often in the form of mathematical models or speculative reviews, are specifically intended to generate hypotheses and predictions and may make no attempt to test them themselves. At times, certain research areas can become overburdened with hypotheses and predictions, generating more than people are able or have the inclination to test. If this happens, it can have a paralysing effect on the development of research. It is thus important that hypotheses, predictions and tests proceed as nearly as is feasible hand in hand.

WHAT THIS BOOK IS ABOUT

Okay, we've said a little about how science works and how the kind of question-asking on which it is based can arise. We now need to look at each part of the process in detail because while each may seem straightforward in principle, some knotty problems can arise when it is put into practice. In what follows, we shall see how to:

1. frame hypotheses and predictions from preliminary source material;

2. design experiments and observations to test predictions;

3. analyse the results of tests to see whether they show anything interesting; and

4. present the results and conclusions of tests so that they are clear and informative.

The discussion deals with these aspects in order so that the book can be read straight through or dipped into for particular points. A summary at the end of each section highlights the important take-home messages and the questions at the end show what you should be able to tackle after reading the book.

Remember, the book is about asking and answering questions in biology – it is not a biology textbook or a statistics manual and none of the points it makes are restricted to the examples that illustrate them. At every stage you should be asking yourself how what it says might apply in other biological contexts, especially if you have an interest in investigating them!

References

Clarke, C. (1990) Professor Sir Ronald Fisher FRS. *British Medical Journal* **301**, 1446–1448.

Verma, G. K. and Beard, R. M. (1981) *What is educational research? Perspectives on techniques of research.* Gower, Aldershot.

2

ASKING QUESTIONS

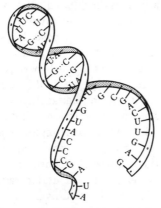

So far, we've discussed asking questions in a very general way. Simply being told that science proceeds like this, however, is not particularly helpful unless it is clear how the principles can be applied to the situation in hand. The idea of this section is thus to look at the development of the procedure in the context of various investigations that you might undertake in practical and project work. We shall assume for the moment that the material of interest is derived from your own observations. We shall start, therefore, with the problem of making observations and directing them in order to produce useful information.

OBSERVATION

Observational notes and measurements

When first confronted with an unfamiliar system, it is often difficult to discern anything of interest straight away. This seems to be true regardless of the complexity of the system. For instance, a common cause of early despair among students watching animals in a tank or arena for the first time is the mêlée of ceaselessly changing activities, many of which seem directionless and without obvious goal. An equally common complaint is that the animals seem to be doing nothing at all worth mentioning. It is not unusual for **both** extremes to be generated by the same animals.

In both of the above cases, the problem almost always turns out to be not what the animals are or are not doing, but the ability of the observer to observe. This is because observation involves more than just staring passively at material on the assumption that if anything about it is interesting then it will also be obvious. To be revealing, observations may need to be very systematic, perhaps involving manipulations of the material to see what happens. They may involve measurements of some kind since some things may be apparent quantitatively rather than qualitatively. In themselves, therefore, observations are likely to involve a certain amount of question-asking. Their ultimate purpose, however, is to provide the wherewithal to frame testable hypotheses and the discriminating predictions that will distinguish between the hypotheses.

To see how the process works, we shall first give some examples of observational notes and then look at the way these can be used to derive hypotheses and predictions. These main examples therefore develop through the book from initial observational notes through framing and testing hypotheses to producing a finished written report. All the examples are based on notes made during practical exercises by our first year undergraduates in biological sciences in their first week at university. In all cases, the observations were made 'cold' without any prior experience or background information about the material. The examples also happen to be behavioural. This is deliberate. In our experience, behaviour provides extensive opportunities for quick experimental manipulation and different forms of analysis (see later) and material for open-ended investigations is cheap and easy to provide for large numbers of students. We therefore actively recommend behaviour as a convenient and undemanding (in terms of cost and previous experience) vehicle for introducing self-driven investigation. The fact that the main examples are behavioural is of course, irrelevant to the aim of the book as the range of other examples running through it amply illustrates. What the examples demonstrate applies with equal weight in all branches of biology from molecular genetics and cell biology to ecology and comparative anatomy.

Example 1

Material: Separate aquaria of 8–10 male and similar number of female guppies (*Poecilia reticulata*), a small aquarium containing water and divided in half with a removable clear or opaque partition, a small handnet.

Notes: First obvious thing is that male and female guppies look very different. Males much more brightly coloured and have longer tails, though females have some bright coloration. Not a lot going on in the tanks, most fish just hanging in the water or moving slowly around. Quite a lot of variation between fish of each sex, especially obvious among males. Some males predominantly red, some have a lot of yellow, others more orange. Vary in size too, and in relative size of body and tail. Some females similar in colour to some males. Measure some fish by viewing against graph paper under the divided tank – males range from 28 to 36 mm (tip of head to tip of tail), females from 30 to 39 mm. Estimate male tail sizes as proportion of total body plus tail size (3 males, tail is 29% of total; 2 males, tail is 33%; 3 males, tail is 25%; 1 male, tail is 37%).

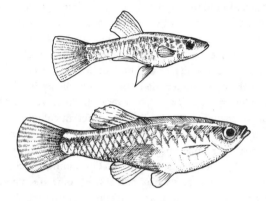

What happens if males and females are put together? Put a male on one side of an opaque partition in the divided tank and a female in the other. Leave them for a few minutes then take away the opaque screen leaving a clear partition. Before the opaque partition removed, both male and female just drift about, nothing obvious happening. When partition removed, male's behaviour changes a lot, apparently casual drifting changes to odd-looking jerky motion with body bent into S-shape. Small pointed fins below at the back extended. Change in behaviour particularly obvious when female close to partition. Male showed postural changes four times in 5 min. Male's colours and markings seem to change too. Female swims up to the partition then moves away – eight approaches by female within 5 min. Try another female – six approaches in 5 min. Change the male for another – this one has more yellow compared with the first (mainly red). Same changes seen in male behaviour (S-shaped posture six times in 5 min); female approaches partition three times in 5 min. Try another female – five approaches to partition in 5 min. Try three more pairs of males and females, males respond as before but only two of the females approach the partition.

Put five males in together. Some males show similar changes to before when female in view but perform behaviours more vigorously and approach closer to the partition. Males showing S-shaped posture did so 6, 12, 11 and 9 times in 5 min. When add more females as well, males appear to direct their attention to some females but not others.

Material: 30–40 1-day-old chicks (*Gallus gallus*), 60 × 60 cm arenas with removable bases, selection of patterned arena bases (e.g. green/orange and red/yellow chequerboard; orange, green, red and yellow painted

Example 2

stones on same colour board), selection of dyed rice grains (orange, green, red and yellow), bench lamps.

Notes: Introduce chick into one of the arenas with a red/yellow chequerboard base and green grain scattered about. Most noticeable thing is loud cheeping, much louder and faster than cheeping from rest of chicks in the stock box. Wanders around arena cheeping. Ignores rice grains. Introduce two more chicks. All three bunch together. Overall cheeping rate seems to drop and cheeps quieter. One chick wanders around the arena and pecks at rice grains and droppings. Other chicks move over to where it pecks and peck at grains and base of arena. Turn on bench lamp over arena. Chicks move under the light and huddle again. Turn off the light and put in four more chicks. Now several chicks move around pecking at rice grains. No cheeping. Throw in some different coloured rice grains. Count number of each colour pecked in 5 min (15 red, 8 green, 0 yellow, 2 orange). Repeat four times (numbers pecked: red 8, 6, 10, 12; green 2, 4, 8, 2; yellow 6, 3, 0, 4; orange 3, 5, 1, 1). Put another chick on its own into the arena with orange stony base and scatter some green grain. Chick sits and cheeps. Remains sitting and cheeping for 10 min. Add another chick. The two huddle then move separately around the arena. One pecks a rice grain then pecks a stone, the other comes over and pecks a stone near the first chick. Count number of pecks/min by any chick over next 5 min (min 1: 4 grains, 4 stones, 2 droppings; min 2: 8 grains, 2 stones; min 3: 6 grains; min 4: 10 grains, 1 stone; min 5: 12 grains, 1 dropping). Remove chicks, clean out arena and scatter orange grain. Put in two different chicks, wait until they start to wander around then count pecks as before (min 1: 1 grain, 3 stones; min 2: 2 grains, 1 dropping; min 3: 2 grains; min 4: 3 grains; min 5: 6 grains).

Example 3

Material: Stock tank of 20–30 three-spined sticklebacks (*Gasterosteus aculeatus*), 30-cm test aquarium of water and gravel, three beakers each containing a different size of *Daphnia*, a small handnet, disposable pipettes.

Notes: Pipette a number of medium-sized *Daphnia* into the test aquarium and allow to disperse. Introduce two sticklebacks into the tank and measure how long it takes for the first *Daphnia* to be eaten (21.3 s). After initial swimming about, fish seem to spend appreciable period hanging motionless in the water between brief, slow movements. Measure periods when motionless over 10 min (6 s, 2 s, 10 s, 4 s, 8 s, 11 s). Take out and introduce four different fish. Time taken to eat first *Daphnia*, 16 s. More of time in tank spent swimming about investigating sides and gravel and taking *Daphnia* (times spent motionless: 4 s, 2 s, 3 s, 8 s, 3 s, 6 s, 4 s). Try a single fish. Swims rapidly round tank then remains motionless near one corner, occasionally swims slowly about then remains motionless again (times spent motionless: 14 s, 3 s, 3 s, 12 s, 2 s, 21 s). Takes 42.5 s to eat first *Daphnia*. Repeat with 10 fish (times spent motionless: 4 s, 3 s, 6 s, 10 s, 2 s, 2 s, 5 s). Takes 8 s to eat first *Daphnia*.

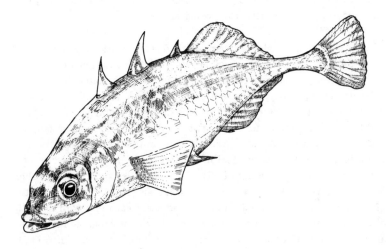

Remove fish and clear tank of *Daphnia*. Introduce mixture of different-sized *Daphnia* into tank and allow to disperse. Introduce stickleback and allow to feed for 5 min. Monitor which *Daphnia* it takes. Fish takes 1 large *Daphnia*, 4 medium-sized ones and 1 small one. Repeat with a second fish – takes 4 small ones. Three more fish take 2 medium-sized and 1 small, 5 small, 4 medium-sized respectively. Introduce a particularly large fish – takes 6 large and 4 medium-sized *Daphnia*.

Example 4

Material: Stock cage of virgin female and stock cage of virgin male field crickets (*Gryllus bimaculatus*), two or three 30 × 30 cm glass/Perspex arenas with silver sand substrate, dish of water-soaked cottonwool and rodent pellets, empty egg boxes, assorted colours of enamel paint, fine paint brush, paint thinners, rule, bench lamps.

Notes: Note that females are distinguished from males by possession of long, thin ovipositor at the back. Put four males into an arena. After rushing about, males move more slowly around the arena. When they meet, various kinds of interaction occur. Interactions involve variety of behaviours: loud chirping, tapping each other with antennae, wrestling and biting. Interactions tend to start with chirping and antenna tapping and only later progress to fighting. Count number of encounters that result in fighting (15 out of 21). Put in three more males so seven in total and count fights again (8 out of 25). Take out males and choose another five. Take various measurements from each male (length from jaws to tip of abdomen, width of thorax, weight) and mark each one with a small, different-coloured dot of paint. Introduce individually marked males into arena. Count number of fights initiated by each male per encounter (red, thorax width 6.5 mm – 4/10; blue, width 7.5 mm – 8/11; yellow, width 7.0 mm – 8/10; silver, 6.0 mm – 3/12; green, 6.5 mm – 3/9). Continue observations and count number of encounters won by each male (win decided if opponent backs off) (red, 3 wins/6 encounters; blue, 4/5; yellow, 7/7; silver, 1/4; green, 4/8).

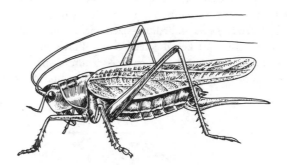

Introduce three sections of egg box to arena and leave males for 10 min. After 10 min, some males (blue, yellow, red) hiding under egg box shelters or sitting close to them. Males in or near boxes chirping frequently. Count number of encounters resulting in fight (12 out of 16). Introduce two females. Females move around the arena but end up mainly around the box sections. Show interest in males and occasionally mount. Other males sometimes interfere when female mounting particular male. Number of approaches to, and number of mounts with, different males: red, 2 approaches, no mounts; blue, 2 approaches, 2 mounts; yellow, 3 approaches, 2 mounts; silver, no approaches; green, 1 approach, 1 mount.

EXPLORATORY ANALYSIS

Observational notes are, in most cases, an essential first step in attempting to investigate material for the first time. However, as the examples amply demonstrate, they are a tedious read and, as they stand, do not make it easy to formulate hypotheses and design more informative investigations. What we need is some way of distilling the useful information so that points of interest become more apparent. If we have some numbers to play with – and this underlines the usefulnesss of making a few measurements at the outset – we can perform some exploratory analyses.

Exploratory analysis may involve drawing some simple diagrams or plotting a few numbers on a scattergram or it might involve calculating some **summary** or **descriptive statistics**. We shall look at both shortly using information from the various sets of observational notes. These sorts of analyses almost always repay the small effort demanded but, like much basic good practice in any field, they are often the first casualty of impatience or prejudgement of what is interesting or to be expected. It is always difficult to discern pattern simply by 'eyeballing' raw numbers and the more numbers we have the more difficult it becomes. A simple visual representation like a scattergram or a bar chart, however, can turn the obscure into the obvious.

Visual exploratory analysis

There are several instances in the examples of observational notes where similar measurements were made under different conditions or when different things were available to animals. For instance, measurements were made of the number of pecks delivered by chicks to different coloured grains of rice and the amount of time sticklebacks spent motionless in the water was measured when different numbers of fish were present. In both of these cases, there are differences in the numbers recorded for different categories (colour of grains) or under different conditions (number of fish). What do these differences suggest?

Eyeballing the rice grain data suggests that some colours are pecked more often than others. A simple way to visualize this might be to total up the number of pecks recorded for each colour across all the sets of observations and present them in a bar chart (Fig. 2.1). If we do this, it looks as though the red grains get more pecks than any of the others. These are followed by green then, some way behind but with almost equal weighting, yellow and orange. From this, we might be tempted to suggest some colour preference among the chicks with red being the hot favourite. However, we shall see later why we need to be a little cautious in our speculation.

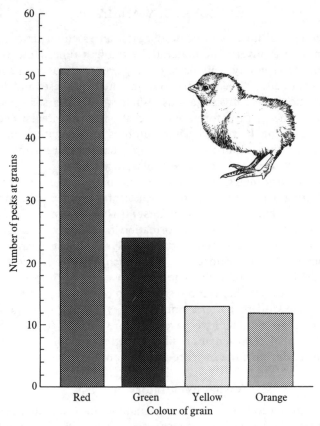

Fig. 2.1

In the case of the fish, we can see that the amount of time individuals spend still in the water seems to change with the number of fish present. An obvious way to see how is to plot a scattergram of time spent motionless against the number of fish in the aquarium. Because the observer did not record the same number of periods of motionlessness for each group size, it would make no sense to total up the times in each case. Instead, we could plot each time independently. The result (Figure 2.2) suggests that, while there is a considerable spread of values for each number of fish, there is a tendency for more time to be spent motionless when fewer fish are present.

These are just two examples. We could do similar things with a number of other measurements in the observational notes. Figure 2.3, for instance, suggests that the number of rice grains pecked per minute by chicks increases with time but that peck rates are generally higher when birds are given green grain. Figure 2.4, on the other hand, suggests that female guppies make

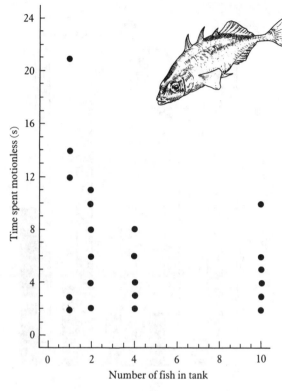

Fig. 2.2

more approaches towards red males than yellow males, at least when males are presented singly. The notes provide scope for other exploratory plots; try some for yourself.

Casting data in the form of figures like this is helpful not just because visual images are generally easier for most people to assimilate than numbers but because they can expose subtleties in the data that are less apparent as raw numbers. Thus even with tiny data sets like those in the examples (and those likely to be generated during brief exploratory periods in classes), points of interest can be highlighted. The plot of time spent motionless in Fig. 2.2, for instance, suggests that, while the amount of time tends to decrease with increasing numbers of fish, the drop from single fish to four fish appears to be greater than the drop from four fish to ten fish. We shall see later that this can lead to some interesting suggestions about the effects of group size on individual behaviour.

The sort of plots we have used so far are useful as a means of seeing at a glance whether something interesting and worthy of further investigation might be going on. However, the data could be presented in a rather different way to make the figures more

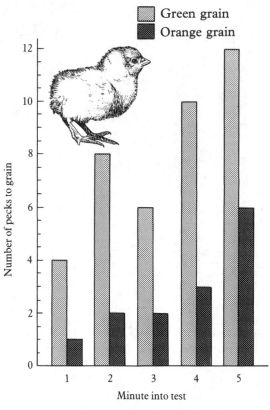

Fig. 2.3

informative. It is clear from both the scattergrams (Fig. 2.2) and the data incorporated into the totals in the bar charts (Figs 2.1, 2.3, 2.4) that there is considerable variability in the numbers recorded in each case; chicks given a choice of different coloured rice grains, for instance, don't all make the same number of pecks at any given colour. This variability has at least two important consequences as far as the exploratory plots are concerned. First, it suggests that simply plotting totals in Figs 2.1 and 2.4 is likely to be misleading if the totals are made up from a wide range of numbers (a large total could be due to a single large result with all the rest being smaller than the results making up other, lesser totals, in which case our interest in the apparent differences between bars might diminish somewhat). Second, variability in data might obscure some potentially interesting tendencies in scattergrams. What we need, therefore, is a way of summarizing data so that (a) the interesting features are still made clear but (b) the all-important variability is also presented, though in a way that clarifies rather than obscures patterns in the data. In short, we need some **summary statistics**.

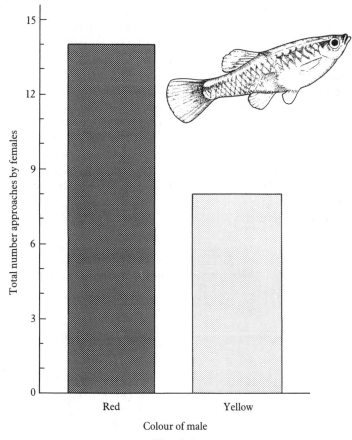

Fig. 2.4

Summary statistics

The usual way of summarizing a set of data so as to achieve (a) and (b) above is to calculate a **mean** (average) or **median** value and then to provide as a measure of the variability the associated **standard error** (for a mean) or **confidence limits** (for a median).

Means and standard errors

The mean (often represented by $\bar{x}$ ('x-bar')) is simply the sum of all the individual values in the data set divided by the number of values (usually referred to as n, the sample size). Formally, the mean is expressed as:

$$\bar{x} = (1/n) \sum_{i=1}^{i=n} x_i$$

The expression $\sum_{i=1}^{i=n} x_i$ indicates that the first $(i = 1)$ to the nth $(i = n)$ data values (x) are summed ($\sum$ is the summation sign) and can be expressed more simply as $\sum x$. This summed value is then multiplied by $1/n$ (equivalent to dividing by the sample size n).

Since the mean is calculated from a number of values, we need to know how much confidence we can have in it. By 'confidence' we mean the reliability with which we could take any such set of values from the material and still end up with the same mean. A statistician would phrase this in terms of our sample mean ($\bar{x}$, the one we have calculated) reflecting the true mean (usually denoted μ) of the population from which the data values were taken. Suppose, for instance, we measured the body lengths of 10 locusts caught in each of two different geographical areas and obtained the following results:

Length (cm)	
Area 1	Area 2
6.3	6.0
7.1	6.4
6.2	6.3
6.5	6.0
7.0	5.9
6.7	6.5
6.5	6.1
7.0	6.2
6.8	6.2
7.1	6.4
$\bar{x}$ 6.7	6.2

Suppose instead, however, that we had obtained the following:

Length (cm)	
Area 1	Area 2
8.3	8.0
5.1	5.4
7.2	5.3
5.5	6.5
8.0	8.4
5.7	5.5
5.5	7.1
8.0	5.2
8.8	5.2
5.1	5.4
$\bar{x}$ 6.7	6.2

In both cases, the mean body lengths from each area are the same and we might want to infer that there is some difference in body size between areas. In the first case, the range of values from which each mean is derived is fairly narrow but different between areas. We might thus be reasonably happy with our inference. In the second case, however, the values vary widely within areas and there is considerable overlap between them. Now we might want to be more cautious about accepting the means as representative of the different areas. We can see that this is the case from the columns of numbers but we need some way of summarizing it without having to present raw numbers all the time. We can do this in several ways. The most usual is to calculate the standard error, a quantitative estimate of the faith that can be placed in the mean and which can be presented with it as a single number. The calculation is straightforward and is shown in Box 1:

How to calculate the standard error of a mean   Box 1

1. Calculate the sum of your data values $(\sum x)$.
2. Square all the data values and sum them, giving $(\sum x^2)$.
3. Calculate $\sum x^2 - (\sum x)^2/n$ (remember n is the sample size, the number of values in your data set). This is actually a quick way of calculating the sum of squared deviations from the mean: $\sum (x - \bar{x})^2$. The deviations are squared so that positive and negative values do not simply cancel each other out.
4. Divide by $n - 1$.
5. Divide by n.
6. Take the square root of the result ...

... and you have the standard error (often denoted as s.e.).

Most scientific calculators will give you the mean of a set of numbers and most will also give you something known as the **standard deviation**, usually represented as σ or s. If your calculator has both σ and σ_{n-1} buttons, it is the σ_{n-1} one that you want. The standard deviation is a different measure that is not important here, but we can use it to obtain the standard error. All we need to do is call up the standard deviation, square it, divide it by n and take the square root.

Whichever way you calculate the standard error (by hand or by calculator), it should be presented with the mean as follows:

$$\bar{x} \pm \text{s.e.}$$

The $\pm$ sign indicates that the standard error extends to its value on either side of the mean. The bigger the standard error, therefore, the more chance there is that the true mean is actually greater or smaller than the mean we've calculated. We can see how this works with our locust data. Let's look at the two sets of values for Area 1, first calculating the x^2 value of the first example:

Locust	Body length (x)	x^2
1	6.3	39.69
2	7.1	50.41
3	6.2	38.44
4	6.5	42.25
5	7.0	49.00
6	6.7	44.89
7	6.5	42.25
8	7.0	49.00
9	6.8	46.24
10	7.1	50.41
$n = 10$	$\sum x = 67.2$	$\sum x^2 = 452.58$

The steps are then:

1. $\sum x = 67.2$

2. $\sum x^2 = 452.58$

3. $\sum x^2 - (\sum x)^2/n = 452.58 - (67.20)^2/10$
$$= 452.58 - 451.58$$
$$= 1$$

4. Divide by $n - 1 = 1/9 = 0.11$

5. Divide by $n = 0.11/10 = 0.01$

6. Take the square root $= \sqrt{0.01} = 0.11$

Thus the mean length of locusts in the first example for Area 1 is:

$$6.72 \pm 0.11 \text{ cm}$$

If we repeat the exercise for the second example, however:

1. $\sum x = 67.2$

2. $\sum x^2 = 471.18$

3. $\sum x^2 - (\sum x)^2/n = 471.18 - 451.58 = 19.6$

4.   Divide by $n - 1 = 19.6/9 = 2.2$

5.   Divide by $n = 2.2/10 = 0.22$

6.   Square root $= \sqrt{0.22} = 0.47$

The mean is now expressed as:

$$6.72 \pm 0.47 \text{ cm}$$

We could leave the mean and standard error expressed numerically like this, or we could present them visually in a bar chart. If we opt for the bar chart, then the mean can be plotted as a bar and the standard error as a line through the top centre of the bar extending the appropriate distance (the value of the standard error) above and below the mean. Figure 2.5a, b shows such a plot for the two sets of example locust data.

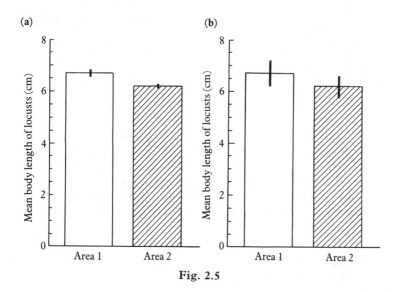

Fig. 2.5

Medians and confidence limits

An alternative summary statistic we could have used is the **median**. There are good statistical reasons, to do with the distribution of data values (see later), why we may need to be cautious about using means. The use of mean values makes important assumptions about the data from which they are calculated that may not hold in many cases. Using medians avoids these assumptions. Later, we shall see that statistical tests of significance can also avoid them. Finding the median is simple. All we do is look for the central value in our data. Thus, if our data comprised the following values:

5   7   11   21   8   12   14

we first rank them in order of increasing size:

5   7   8   11   12   14   21

and take the value that ends up in the middle, in this case 11. If we have an even number of values so there isn't a single central value, we take the halfway point between the two central values, thus in the following:

2   6   8   20   23   38   40   85

the median is 21.5 (halfway between 20 and 23). Note that the median may yield a value close to or very different from the mean. In the first example, the mean is 11.1 and thus similar to the median. In the second sample, however, it is much greater at 27.58.

Again, we want some way of indicating how much confidence to place in the median. By far the simplest way is to find the **confidence limits** to the median using a standard table, part of which is shown in Appendix I. All we need to do is rank order our data values as before, count the number of values in the sample (n), then use n to read off a value r from the table. Normally, we would be interested in the r value appropriate to confidence limits of approximately 95% ('approximately' because, of course, the limits always have to be two of the values in the data set – if n is less than 6, 95% confidence limits cannot be found). This r value then dictates the number of values in from the two extremes of the data set that denote the confidence limits. Thus in our first sample data set, there are seven values. Reference to Appendix I shows that for $n = 7$, $r = 1$; the confidence limits to the median of 11 are therefore 5 and 21. However, if we had a sample of nine values (say 7, 11, 15, 22, 46, 67, 71, 82, 100) r for approximately 95% confidence limits is 2, so for a median of 46, the limits would be 11 and 82.

As with the mean and standard error, we can represent medians and their associated confidence limits visually in a bar chart.

Exploratory analysis, then, is a way of summarizing data, visually or numerically, to make it easier to pick out interesting features. This is useful, however, only to the extent that it promotes further investigation to confirm that what looks interesting at an exploratory level is still interesting when data are collected more rigorously and subjected to more thorough analysis. This brings us back to hypotheses and predictions and leads to a consideration of **confirmatory statistics**.

FORMING HYPOTHESES

Turning exploratory analyses into hypotheses and predictions

Exploratory analyses are generally open-ended in that they are not guided by preconceived ideas about what might be going on. However, they are the first important step on the way to formulating hypotheses which do then guide investigation. As we have seen already, hypotheses can be very general or they can be specific. Both kinds can be generated from our observational notes.

The observations of male and female guppies suggest fairly strongly that the odd posturing and associated behaviour by males is concerned with attracting females. When males and females were allowed to see each other through a partition, males performed the striking behaviour and females showed some tendency to approach them. A testable hypothesis is thus:

Example 1

Guppies

Hypothesis 1A *The S-shaped male posture is a courtship display designed to attract females.*

and some predictions might be:

Prediction 1A(i) *Males will display only when a female is in view.*

Prediction 1A(ii) *Males will display more vigorously to larger females* (on the assumption that larger females carry more eggs).

Prediction 1A(iii) *The attractiveness of males will increase with display vigour and females will show a greater tendency to approach males that display more vigorously.*

Hypothesis 1A is a very broad hypothesis and could give rise to a large number of predictions about interactions between males and females. The observational notes, however, also suggest some more specific hypotheses

that could be tested. For example:

Observation: Females approached red males more often than yellow males.

Hypothesis 1B *The colour of a male influences his attractiveness to females.*

Prediction 1B *Given a choice of different coloured males, females will approach males of some colours more than those of other colours for any given level of display vigour on the part of the male.*

Observation: Males performed S-shaped displays more often in groups.

Hypothesis 1C *Competition with other males means males must display more vigorously to attract females.*

Prediction 1C *Individual display vigour will increase with the number of males present.*

Example 2

Chicks

The observations of chicks focused on various aspects of the birds' behaviour in arenas, some to do with social interaction, others with feeding. Several hypotheses can thus be derived from the notes. Some of these have already become apparent in the discussion of exploratory analysis.

Observation: Cheeping is louder and more frequent when chicks are removed from the stock box to an arena.

Hypothesis 2A *Cheeping is a contact call used to maintain proximity to companions.*

Prediction 2A *Chicks will cheep more when they feel isolated.*

Observation: Chicks move over to where another has just pecked at the ground.

Hypothesis 2B *Other chicks provide information about the form and location of food.*

Prediction 2B *Chicks will find food faster if they forage with companions.*

Observation: Chicks pecked red grain more than other colours when given an arbitrary mixture of colours.

Hypothesis 2C *Chicks show colour preferences when choosing food.*

Prediction 2C *Given a choice of different coloured food items, chicks will take more items of certain colours per encounter than others.*

Observation: Peck rates to orange grains on a green background were higher than to green grains on the same background.

Hypothesis 2D *The conspicuousness of novel food items affects the ability of chicks to see them.*

Prediction 2D *The rate at which chicks peck at food items will be greater when the items are presented on a different coloured background rather than on a similar coloured background.*

Observations of three-spined sticklebacks suggested that the amount of time fish spent in different behaviours varied with the number of individuals present. Fish also seemed to differ in the number and size range of *Daphnia* eaten.

Example 3

Sticklebacks

Observation: The amount of time spent motionless in the water tended to decrease when more fish were present.

Hypothesis 3A *Lack of movement reflects vigilance for predators. Fish feel safer in groups so spend less time vigilant.*

Prediction 3A *Time spent motionless by individuals will decrease with increasing numbers of fish but will be greater for any given number if fish are alarmed.*

Observation: The time taken to eat the first *Daphnia* tended to be shorter when more fish were present.

Hypothesis 3B *The reduced vigilance in larger groups allows more time to look for food.*

Prediction 3B *Individual feeding rate will increase with the number of fish.*

Observation: A large fish ate more large *Daphnia* than the other fish observed.

Hypothesis 3C *Body size affects the size range of prey that are most profitable* (*result in the greatest net rate of energy return per unit time spent eating*) *to the fish.*

Prediction 3C *The profitability of different sized* Daphnia *will correlate with the size of the fish and, given a choice, fish will take predominantly the most profitable sizes.*

Example 4

Crickets

Male field crickets seemed to be aggressive to one another when put together in an arena. Whether or not an encounter resulted in fighting varied with the number of crickets and individuals differed in their tendency to initiate and win fights. The apparent effects of providing egg box shelters and introducing females suggest that interactions between males are concerned ultimately with gaining access to females.

Observation: The number of encounters leading to a fight was lower when more crickets were present.

Hypothesis 4A *The cost of fighting on encounter increases with population size and the chance of encountering another male.*

Prediction 4A *The probability of an encounter resulting in a fight will decrease with increasing numbers of males and in the same number of males maintained at a higher density.*

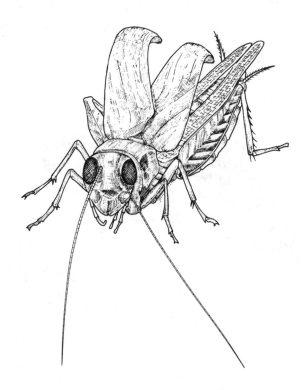

Observation: Larger males initiated more fights per encounter and won in a greater proportion of encounters.

Hypothesis 4B *Large size confers an advantage in fights between males.*

Prediction 4B *Males will be less likely to initiate a fight when their opponent is larger.*

Observation: Interactions tended to escalate from chirping and antenna-tapping to overt fighting.

Hypothesis 4C *The escalating sequence reflects information gathering regarding the size of the opponent and the likelihood of winning.*

Prediction 4C *Encounters will progress further when opponents are more similar in size and it is more difficult to judge which will win.*

Observation: Larger males ended up in or near egg box shelters and females tended to spend more time with these males.

Hypothesis 4D(i) *Females prefer to mate with males in shelters for protection from predators.*

Prediction 4D(i) *Giving a male a shelter will increase the attention paid to him by females and his chances of copulating.*

Hypothesis 4D(ii) *Females prefer large males.*

Prediction 4D(ii) *Given a choice of males, all with or without shelters, females will spend more time and be more likely to copulate with larger males.*

Null hypotheses

In the examples above, we have phrased predictions in terms of the outcome they lead us to expect. Prediction 1C, for example, leads us to expect that male guppies will display more vigorously in groups. We can then test the prediction by doing a suitable experiment and carrying out some confirmatory analysis. Formally, however, we do not test predictions in this form. Rather, we test them in a null form that is expressed as a hypothesis **against** the prediction. This is known as a **null hypothesis** and is often expressed in shorthand as H_0. Predictions are tested in the form of a null hypothesis because science proceeds conservatively, always assuming that something interesting is **not** happening unless convincing evidence suggests, for the moment, that it might be. In the case of Prediction 1C, therefore, the null hypothesis would be that increasing the number of males makes no difference to the vigour of their displays. We shall see later what burden of proof is necessary to enable us to reject the null hypothesis in any particular case.

There is a second, and from a practical point of view more crucial, point to make about the predictions. Skimming down them gives the impression of specificity and diversity; each prediction is tailored to particular animals and circumstances and those from one example seem to have little to do with those from others. At the trivial level of detail, this is obviously true. However, in terms of the kinds of question they reflect, predictions from the different examples in fact have a great deal in common. Before we can proceed with the problem of testing hypotheses and choosing confirmatory analyses, we need to be aware of what these common features are.

Differences and trends

Although we derived some 17 different predictions from our notes, and could have derived many more, all fall (and would have fallen) into one of two classes. Regardless of whether they are concerned with guppies or crickets or with finding food or mate attraction, they either predict some kind of **difference** or they predict some kind of **trend**. Recognizing this distinction is vitally important because it determines the kind of confirmatory test we shall be looking to perform and therefore the design of our experiments. Surprisingly, however, it proves a stubborn problem for many students throughout their course with the result that confirmatory analyses often fall at the first fence. Let's look at the distinction more closely.

A **difference** prediction is concerned with some kind of difference between two or more groups of measurements. The groups could be based on any characteristics that can be used to make a clear-cut distinction; obvious examples could be sex (e.g. a difference in body size between males (Group 1) and females (Group 2)), functional anatomy (e.g. a difference in enzyme activity between xylem (Group 1), phloem (Group 2) and parenchyme (Group 3) cells in the stem of a flowering plant) or experimental treatment (e.g. a difference in the number of chromosomal abnormalities following exposure to a mutagen (Group 1) or exposure to a harmless control (Group 2)). Which of the predictions we derived earlier are difference predictions?

Example 1

Guppies

In the guppy example, there are two difference predictions:

Prediction 1A(i) leads us to expect a difference in the vigour of male display between female present (Group 1) and female absent (Group 2) treatments.

Prediction 1B leads us to expect a difference in the number of approaches by females to males of different colour (e.g. Group 1, red males; Group 2, yellow males; Group 3, blue males).

All the predictions arising from observations of chicks turn out to be difference predictions:

Example 2

Chicks

From Prediction 2A, we expect a difference in the amount of cheeping between isolated chicks (Group 1) and chicks in a group (Group 2).

Prediction 2B is also concerned with a difference between isolated chicks (Group 1) and chicks in groups (Group 2), but this time in the speed of finding food.

Prediction 2C suggests a difference between food colours in the amount taken by chicks (e.g. Group 1, orange food; Group 2, green food; Group 3, yellow food; Group 4, red food).

Prediction 2D again deals with feeding behaviour, this time predicting a difference in the amount of a given coloured food taken on the same colour background (Group 1) versus a different colour background (Group 2).

Part of two predictions in the stickleback examples involves differences:

Example 3

Sticklebacks

Part of Prediction 3A leads us to expect a difference in time spent motionless between fish that have been alarmed (Group 1) and fish that have not been alarmed (Group 2).

Part of Prediction 3C suggests that, when presented with a range of sizes of *Daphnia*, fish will take mainly the most profitable size. The groups are thus based on prey size (e.g. Group 1, small *Daphnia*; Group 2, medium-sized *Daphnia*; Group 3, large *Daphnia*).

Two predictions from the crickets involve differences:

Example 4

Crickets

Prediction 4B suggests that males will be less likely to initiate a fight when their opponent is larger than them (Group 1) than when it is smaller (Group 2).

Prediction 4D(i) predicts that females will pay more attention to males with a shelter (Group 1) than to males without a shelter (Group 2).

Trend predictions are concerned not with differences between hard and fast groupings but with the relationship between two more or less continuously distributed measures. Thus, for example, a relationship might be predicted between the amount of an anthelminthic drug administered to a rat infected with nematodes and the number of worm eggs subsequently counted in the animal's faeces. In this case, we should expect the relationship to be negative with egg counts decreasing the more drug the rat has received. On

the other hand, a positive relationship might be predicted between the number of hours of sunlight received and the standing crop of a particular plant. With trends we can therefore envisage two measures as the axes of a graph. One measure extends along the bottom (x) axis, the other up the vertical (y) axis. Sometimes it doesn't matter which measure goes along the x-axis and which up the y-axis because there is no basis for implying cause and effect and we are interested only in whether there is some kind of association. Thus, we might expect a strong association between the amount of ice cream eaten and the amount of time spent in the sea on a visit to the seaside because both would go up with temperature. Since neither could reasonably be thought of as a cause of the other, it is of no consequence which goes on the x- or y-axis. In the two examples above, however, there are reasonable grounds for supposing cause and effect. While it is plausible for the anthelminthic drug to affect faecal egg counts, it is not plausible for the egg counts to have influenced the amount of drug. Similarly, hours of sunlight could influence a standing crop but not vice versa. In these cases, the drug dose and hours of sunlight measures should go on the x-axis and the egg counts and standing crops on the y-axis. It is important to stress, however, that by doing this we are not asserting that the x-axis measure really is a cause of the y-axis measure – as we shall see later, inferring cause and effect from relationships requires extreme caution – merely that if there was a cause and effect relationship it would most likely be that way round. This is also clear in the remainder of our example predictions all of which involve trends.

Example 1
Guppies

In prediction 1A(ii), males are expected to display more vigorously as female size increases. Female size should thus be the x measure and male display vigour the y measure.

Prediction 1A(iii) suggests that male attractiveness (measured as approach tendency by females) will increase with display vigour. Now male display vigour becomes the x measure with female approach tendency as the y measure.

In Prediction 1C, male display vigour is once again the y measure with the number of males present as the x measure.

Example 3
Sticklebacks

The first part of Prediction 3A leads us to expect time spent motionless (y) will decrease with the number of fish (x) while Prediction 3B suggests feeding rate (y) will increase with the number of fish (x).

The first part of Prediction 3C predicts a relationship between fish size (x) and the size of *Daphnia* that is most profitable (y).

Prediction 4A first of all predicts that the number of encounters ending up in a fight (y) will increase with the number of crickets (x), then predicts a similar increase when the same number of crickets are maintained at higher densities (density = x).

Example 4

Crickets

Prediction 4C involves a predicted trend in the tendency to escalate an interaction (y) with decreasing difference in size between opponents (x).

Prediction 4D(ii) suggests that the time females spend with a male (y) and their tendency to copulate (y) will increase with male size (x).

There is thus a clear distinction between difference and trend predictions. Of course, it is possible to recast some trend predictions as difference predictions (for instance, a continuous measure of group size for use in a trend could always be recast in terms of small groups (groups below size w) and large groups (groups above size w) and thus be used in a difference prediction). What makes the distinction, therefore, is not the data *per se* but the way data are to be collected or classified for analysis. Thus, while measures such as group size or time intuitively suggest trends, there is nothing to stop them being used in difference predictions. It all depends on what is being asked. This is often a source of serious confusion among students encountering open-ended data-handling for the first time.

SUMMARY

1. Open-ended observation is a good way to develop the basis for forming hypotheses and predictions about material. It pays to make observations quantitative where possible so that exploratory analyses can highlight points of interest.

2. Exploratory analysis is a useful (often essential) first step in extracting interesting information from observational notes or other sources of exploratory information. It can take a wide variety of forms, such as bar charts, scattergrams, or tables of summary statistics.

3. Exploratory analyses, or raw exploratory information itself, can lead to a number of hypotheses about the material. In turn, each hypothesis can give rise to several predictions that test it. Formally, predictions are tested in the form of null hypotheses.

4. While predictions derived from hypotheses may be diverse and specific in detail to the material of interest, they fall into two clearly distinguishable categories: predictions about **differences** and predictions about **trends**. Which of these categories a prediction belongs to is determined by the way data are to be collected or classified for analysis.

3

ANSWERING QUESTIONS

In the last chapter, we looked at the way hypotheses can be derived from exploratory information. We turn now to the problem of how to test our hypotheses. As we have seen, we begin by making predictions about what should be the case if our hypotheses are true. These predictions then dictate the experiments or observations that are required. However, this may not be as straightforward as it sounds; decisions have to be made about what is to be measured and how, and how the resulting data are to be analysed. The questions of measurement and analysis are, of course, interdependent. This is obvious both at the level of choosing between difference and trend analyses – there is little point collecting data suitable for a difference analysis if what we're looking for is a trend – and at the level of analyses **within** differences and trends. We shall come to this later. While at first sight it might seem like putting the cart before the horse, therefore, we shall introduce confirmatory analysis **before** dealing with the collection of data so that the important influence of choice of analysis on data collection can be made clear.

CONFIRMATORY ANALYSIS

The need for a yardstick in confirmatory analysis: statistical significance

Take a look at the scattergram in Fig. 3.1. It shows a relationship between the concentration of a fungicide sprayed on a potato crop and the percentage of leaves sampled subsequently that showed

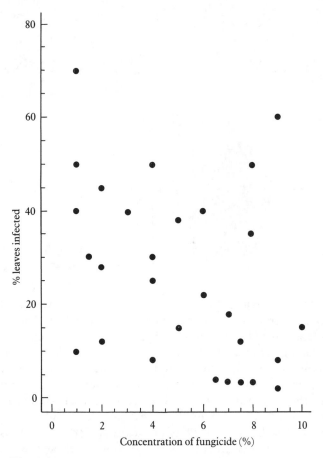

Fig. 3.1

evidence of fungal infection. A plot like this was presented recently to a class of undergraduates. Students in the class were asked whether they thought it suggested any effect of fungicide concentration on infection. The following are some of their replies:

Yes, fungicide concentration obviously has an effect because infection goes down with increasing concentration.

It's hard to say. It looks as though there is some effect, but it's pretty weak. More data are needed.

Fungicide concentration is reducing infection but there must be other things affecting it as well because there's so much scatter.

I don't think you can say anything from this. Yes, there is some downward trend with increasing concentration but several points for high concentrations are higher than some of those for low concentrations. Totally inconclusive.

Yes, there is a clear negative effect.

Clearly there are different, subjective, reactions to the plot. To some it is unequivocal evidence for an effect of fungicide concentration: to others it doesn't suggest much at all. Left to 'eyeball' impressions, therefore, the conclusion that emerged would be highly dependent on who happened to be doing the eyeballing. What is required, quite clearly, is some independent yardstick for deciding whether or not we can conclude anything from the relationship. Since the scenario above could be repeated with any set of data – difference or trend – the need for such a yardstick arises in all cases. One or two idiosyncratic departures notwithstanding (one well-known ornithologist used to advocate the yardsticks 'not obvious', 'obvious' and 'bloody obvious'), the yardstick used conventionally in science is **statistical significance**. There is nothing magic or complicated about statistical significance. It is simply an arbitrary criterion accepted by the international scientific community as the basis for accepting or rejecting the null hypothesis in any given instance and thus deciding whether predictions, and the hypotheses from which they are derived, hold. If the criterion is reached, the difference or trend in question is said to be **significant**; if it is not, the result is **non-significant**. The term 'significant' thus has an important, formal meaning in the context of data analysis and its use in a casual, everyday sense should be avoided in discussions relating to scientific interpretation. How do we decide whether differences or trends are significant? By using the most appropriate of the vast range of significance tests at our disposal. Before we introduce some of these tests, however, we must say a little more about significance itself.

WHAT IS STATISTICAL SIGNIFICANCE?

The criterion that determines significance is the probability (often denoted as α) that a difference or trend as extreme as the one observed would have occurred if the null hypothesis that there is really **no** difference or trend in the population from which the sample came was true. Confused? An example makes it clear. Let's take one of our earlier predictions, say Prediction 3B. Here, we are predicting that individual sticklebacks will feed at a higher rate as group size increases. Suppose we have tested this by measuring the feeding rate of an arbitrarily chosen fish in each of ten different group sizes and obtained what looks like a convincing positive trend: feeding rate goes up with group size. The null hypothesis in this case, of course, is that feeding rate does **not** increase with group size. What, then, is the probability of obtaining a positive relationship as extreme as the one we got if this null hypothesis is really true and the apparent trend a chance effect? A helpful analogy here might be the probability of obtaining the apparent trend by haphazardly throwing darts at the scattergram. An appropriate significance test will tell us (we shall see how later). By convention in biology, a probability of 5% (= 0.05 when expressed as a proportion) is accepted as the threshold of significance. If the probability of obtaining a relationship as extreme as ours by chance turns out to be 5% or less, we can regard the relationship as significant and reject the null hypothesis. If the probability is greater than 5% we do not reject the null hypothesis and the relationship is regarded as non-significant. If the null hypothesis is not rejected, we effectively assume that our apparent relationship was due to a chance sampling effect. As a matter of interest, the negative trend in Fig. 3.1 is significant at the 5% level, so the optimists have it in this case!

The 5% threshold is, of course, arbitrary and still leaves us with a one-in-twenty chance of rejecting the null hypothesis incorrectly (falsely accepting there is a difference or a trend when there isn't). Under some circumstances, for instance when testing the effectiveness of a drug, a one-in-twenty risk of incorrect rejection might be considered too high. In certain areas of research such as medicine, therefore, the arbitrary threshold of significance is set at 1% (= 0.01). In other disciplines it is sometimes relaxed to 10% (= 0.1).

An important point must be made here regarding the inference to be drawn from achieving different levels (10%, 5%, 1%, etc.) of significance. A high level of significance is not the same as a large effect in the sense of a large difference or a steep trend. The magnitude of an effect – difference or trend – is usually known as an **effect size**. This is quite different from the level of significance. The distinction is made clear in Fig. 3.2. The figure shows two trends. In Fig. 3.2a, the y measure increases in close relationship with the x measure yielding what looks like a clear positive trend. Figure 3.2b, on the other hand, shows a scatter of points in which it is more difficult to discern a trend. However, if we perform a suitable significance test for a trend on the two sets of data, the trend in Fig. 3.2b turns out to be significant at the 1% level while that in Fig. 3.2a isn't even significant at the 10% level. The crucial difference between the two trends, of course, is the sample size. The number of data points in Fig. 3.2a is low so a few inconsistencies in the trend are enough to push it

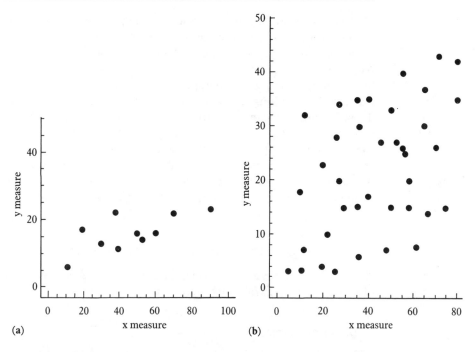

Fig. 3.2

below significance. Figure 3.2b, however, has a large number of points so even though there is a wide scatter, the trend is still significant. Exactly the same sample size effect would operate in the case of difference analyses.

Because the level of significance by itself gives little indication of the magnitude of a difference or trend, it is always helpful to provide such an indication, usually in the form of summary statistics and their associated sample sizes. We shall return to this point later.

SIGNIFICANCE TESTS

So far, we have seen how to get round the problem of subjective impression in interpreting data by using the criterion of statistical significance, and we have looked at some caveats on the interpretation of significance. We can now turn to the statistical **tests** that enable us to decide significance.

Right at the beginning we said that this book was not about statistics. It isn't. At the same time, statistical significance tests are an essential tool in scientific analysis and the rules for using them must be clearly understood. However, this no more demands a knowledge of statistical theory and the mathematical mechanics of tests than using a computer package demands an appreciation of electronics and microcircuitry. As with most tools, it is competent **use** that counts rather than theoretical understanding. Nevertheless, acquaintance with statistical theory is certainly to be encouraged and it is envisaged that many users of this book will also be pursuing courses in statistics and will have at their disposal some of the many introductory and higher-level textbooks now available (e.g. Bailey, 1981; Meddis, 1984; Sokal and Rohlf, 1981). We stress again, though, that this is neither assumed nor required for our purposes. Our aim here is simply to introduce the use of significance testing as a basic tool of enquiry and some simple, but broadly applicable, tests for differences and trends.

Types of measurement and types of test

The test we choose in a particular case may depend on a number of things only three of which need concern us here.

Types of measurement

The first is the kind of measurement we employ. Without getting too bogged down in jargon, we can recognize three kinds.

Nominal or classificatory measurement. Here, observations or recordings are allocated to one of a number of discrete, mutually exclusive categories such as male/female, mature/immature, red/yellow/green/blue, etc. Thus if we were to watch chicks pecking at red (R), green (G) and orange (O) grains of rice and recorded the sequence of pecks with respect to the colour targeted, we might end up with a string of data as follows:

R O R R G G O R G G G G O O G R O

Such data are measured purely at the level of the category to which they belong and measurement is thus **nominal** or **classificatory**.

Ordinal or ranking measurement. In some cases, it may be desirable (or necessary) to make measurements that can be **ranked** along some kind of scale. For instance, the intensity of the colour of a turkey's wattles might be used as a guide to its state of health. The degree of redness of the wattles of different birds could be scored on a scale of 1 (pale pink) to 10 (deep red). The allocation of scores to wattles is arbitrary and there is no reason to suppose that the degree of redness increases by the same amount with each increase in score. Thus the difference in redness between scores of 8 and 9 might be greater than the difference between scores of

2 and 3. All that matters is that 9 is redder than 8 and 3 is redder than 2; the absolute difference between them cannot be quantified meaningfully.

Constant interval measurements.   In other scale measurements, the difference between scores **can** be quantified so that the difference between scores of 2 and 3 is the same as that between scores of 8 and 9. Such measurements may have arbitrarily set (e.g. scales of temperature) or true (e.g. scales of time, weight, length) zero points. Such **constant interval measurements** can in fact be split into two categories on the basis of arbitrary versus true zero points and their scaling properties (e.g. Martin and Bateson, 1986), but this is not important here.

While defining measurements seems rather dry and theoretical, we need to be aware of the kind of measurement we use because, as we shall see, some significance tests are very fussy about the form of data they can accept.

Parametric and nonparametric significance tests

The second thing we must keep an eye on is the nature of the data set itself, in particular the sample size and the distribution of values within the sample. Again, detailed consideration of this is unnecessary but it is a factor that determines the range of tests we shall be introducing so a brief discussion is warranted.

There are two classes of statistical test that differ in the assumptions they make about data. **Parametric** tests make a number of important assumptions that are frequently violated by the kind of data samples collected during practical exercises. The most critical concerns the distribution of values within samples. Parametric tests assume that the data with which they are dealing conform (reasonably closely at any rate) to what is known as a **normal** distribution. This is illustrated in Fig. 3.3. Essentially it demands that most of the data values fall in the middle of the range (cluster about the mean) with the number tapering off symmetrically either side of the mean to a few extreme values in each of the two tails. The height of the adult male or female population in a city would look something like this; most people would be around the average height for their sex, some would be quite tall or quite short and a few would be extremely tall or extremely short. The arithmetic of parametric tests is based on the parameters describing this symmetrical, bell-shaped curve (hence the term 'parametric'). The more distorted (less normal) the distribution becomes, therefore, the less meaning the calculations of parametric tests have. Parametric tests also demand

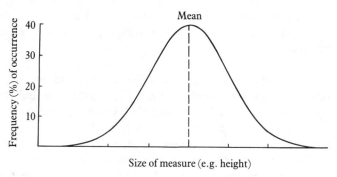

Fig. 3.3

that measurements are of the constant interval kind, so cannot usually deal with the other types of measurement we might be forced to use.

In contrast, **nonparametric** tests make no assumptions about the distribution of data values, although they do assume some other attributes with which, happily, it is often easier to comply. In the statistical jargon, they are thus more **robust** because they are capable of dealing with a much wider range of data sets. Many nonparametric tests work on rank orders of data values. While they can deal with the same constant interval measurements as parametric tests, therefore, they can also cope better with ordinal (ranking) and classificatory measurements. Nonparametric tests are especially useful when sample sizes are small and the assumptions of normality particularly troublesome. However, nonparametric methods are not entirely without shortcomings. They are slightly less powerful ('power' meaning the probability of properly rejecting the null hypothesis) than their parametric equivalents in most cases, and when it comes to more complex analyses involving several variables at the same time, the avail-ability of suitable nonparametric methods is limited, though beginning to expand (see, for example, Meddis, 1984). Never-theless, because of their wide applicability and usually simple calculation, all the tests we shall introduce here, with the exception of one, will be nonparametric.

One-tailed versus two-tailed, and general versus specific tests

The third important factor we must consider relates to the prediction we are trying to test. Suppose we are predicting a difference between two sets of data, say a difference in the rate of growth of a bacterial culture on agar medium containing two different nutrients. We could make two kinds of prediction. On the one hand, we could predict a difference without implying anything about which culture should grow faster. In this case, we

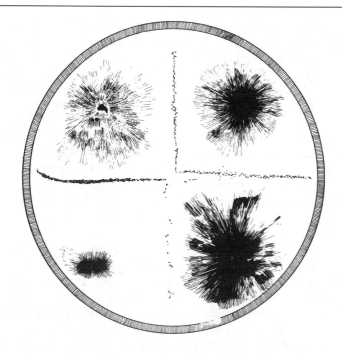

wouldn't care whether culture A grew faster than culture B or
vice versa. This is a **general** prediction. On the other hand, we
might predict that one particular culture would grow faster than
the other, e.g. A would grow faster than B; this is a **specific**
prediction. Which of these kinds of prediction we make affects the
way we test the predictions.

The same distinction arises with trend predictions. Imagine we
want to know whether there is a trend between the size of a male
cricket and the number of fights he wins over the course of a day.
We can make a general prediction (there will be a trend, positive
or negative), or we can make a specific prediction (larger males
will win more fights; i.e. the trend will be positive). We can think
of the general prediction as incorporating both positive and
negative trends; either would be interesting. The specific prediction is concerned with only one of these.

In cases like those above, where there are only two possible
specific predictions within the general one, we can use the same
significance test for either general or specific predictions but with
different threshold probability levels for the test statistic (see
below) to be significant. Because here the specific and general
predictions are concerned with one and two directions of effect
respectively, the threshold value of the test statistic at the 5%
level
in the general test becomes the threshold value at the 10% level
in the specific test. In statisticians' jargon, we thus do either a
one-tailed (specific) or a two-tailed (general) version of the same

test. There is, of course, an obvious, and dangerous, trap for the unwary here. The trap is this: if the value of a test statistic just fails to meet the 5% threshold in a two-tailed test, there is a sore temptation to pretend the test is really one-tailed so that the test statistic value appropriate to the 10% probability level in a two-tailed test can be used, thus increasing the likelihood of achieving a significant result. **It must be stressed that this is tantamount to cheating. A one-tailed test is legitimate only when the prediction is made in advance of obtaining the result. It is completely inadmissible as a fallback when a two-tailed test fails to yield a significant outcome. A one-tailed test should thus be used only when there are genuine** reasons for predicting the direction of a difference or trend **in advance.**

However, the distinction between general and specific tests is not simply that between two-tailed and one-tailed tests. Instead of the above, for instance, imagine that we are predicting a difference between three (or more) groups, say weight of fruit produced in a season from trees treated in three different ways: A, B and C. As before, we can make the general prediction that there will be a difference between the three treatments but, unlike before, this general prediction incorporates **six** potential specific predictions about how the treatments will differ ($A > B > C$, $B > C > A$, $C > A > B$, $A > C > B$, $B > A > C$, $C > B > A$). This extra complexity means that we have to use different tests for a general as opposed to a specific prediction – we shall see how later – and cannot simply use one-tailed and two-tailed versions of the same test. The one-tailed/two-tailed distinction is thus a special case of the difference between general and specific tests.

Simple significance tests for differences and trends

Having discussed the general principle of statistical significance, we come now to some actual tests which allow us to see whether differences or trends are significant at an appropriate level of probability. A glance at any comprehensive statistics textbook will reveal a plethora of difference tests for both kinds of analysis. These cater for the various subtleties of assumption and require-ment for statistical power under different circumstances. Many of these tests, however, are sophistications of more basic tests which are suitable for a wide range of analyses. Here, we introduce a selection of such basic tests which can be used with most kinds of data that are likely to be collected in practical exercises. These are presented in outline in the text but detailed examples of calculations are given in Appendix II. The tests can thus be calculated by hand if necessary.

Test statistic

Significance tests calculate a test statistic that is usually denoted by a letter or symbol: t, H, F, χ^2, r, r_s and U are a few familiar examples from various parametric and nonparametric tests. The value of a test statistic has a known probability of occurring by chance for any given sample size or what are known as 'degrees of freedom' (see later). A calculated value can thus be checked to see whether it concurs with or passes (positively or negatively, depending on the test) the threshold value appropriate to the level of probability chosen for significance. This usually means comparing the value with a table of threshold values. Such comparisons are made automatically for the tests in many statistical computer software programs.

Tests for a difference

We shall introduce three types of difference test: χ^2 (chi-squared, pronounced 'ky-squared'), the Mann–Whitney U test, and analysis of variance. All test predictions of differences, but each deals with a different kind of measurement or number of groups being compared.

Tests for a difference between two groups

We shall start with the relatively simple situation of comparing two groups. Here, two mutually exclusive groups (e.g. male/female, small/large, with property a/without property a, etc.) have been identified and measurements made with respect to each (e.g. the body length of males versus the body length of females, the number of seeds set by small plants versus the number set by large plants, the survival rate of mice on drug A versus the survival rate on drug B). Depending on the kind of measurement made, we can use one of a number of tests to see whether any differences are significant. Two are introduced here.

• *Chi-squared* (χ^2). A chi-squared test can be used if data are in the form of counts, i.e. if two groups have been identified and observations classified in terms of the number belonging to each. Chi-squared can be used **only** on raw counts; it cannot be used on measurements (e.g. length, time, weight, volume) or proportions, percentages, or any other derived values. The test works by comparing observed counts with those expected by chance or on some prior basis. As an example, we can consider a simple experiment in Mendelian inheritance. Suppose we crossed two pea plants that are heterozygous for yellow and green seed colour, with yellow being dominant. Our expectation from the principles of simple Mendelian inheritance, of course, is that the progeny will exhibit a seed colour ratio of 3 yellow:1 green. We can use the chi-squared test to see whether our observed numbers of yellow and green seeds differ from those expected on a 3:1 ratio. The expected numbers are simply the total observed number of progeny divided into a 3:1 ratio. Thus:

	Seed colour		
	Yellow	Green	Total
Number observed	130	46	176
Number expected	132	44	176

To find our χ^2 test statistic we calculate:

$$\chi^2 = \Sigma(O - E)^2/E$$

where O is the observed number and E the expected number in each group. In our example, therefore:

$$\chi^2 = (130 - 132)^2/132 + (4 - 44)^2/44$$

$$= 0.0303 + 0.0909$$

$$= 0.1212$$

We can now check our calculated value of χ^2 against the threshold values in Appendix III, Table A. To do this, however, we must first decide on the appropriate number of **degrees of freedom** (often referred to simply as d.f.) to use. Degrees of freedom are related to sample size and/or the number of groups or classes into which data are cast (how and why need not concern us) and are calculated in a number of different ways depending on the test. In the two-group chi-squared test above, the number of degrees of freedom is equal to the number of groups minus one (i.e. $2 - 1 = 1$). The threshold χ^2 value in which we are interested is therefore the value appropriate to a 5% (0.05) probability for 1 degree of freedom. Reference to Table A shows this value to be 3.84. To be significant at the 5% level, our calculated χ^2 value must **equal or exceed** 3.84. Since 0.1212 is less than 3.84, the null hypothesis that there is no departure from a 3:1 ratio cannot be rejected and there is no reason to suppose that seed colour is segregating in anything other than a simple Mendelian fashion. If we had set our threshold of significance at 1% (0.01) instead of 5%, the value that χ^2 would have had to equal or exceed is 6.63 (Table A). The criterion thus becomes tougher the smaller the margin of error we impose.

In the pea example, the expected numbers were dictated by the Mendelian theory of inheritance; we have good reason to expect

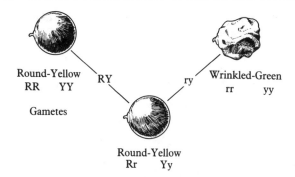

Round-Yellow RY
RR   YY

ry   Wrinkled-Green
rr   yy

Gametes

Round-Yellow
Rr   Yy

a 3:1 ratio of yellow: green seeds and thus a difference between groups. In many cases, of course, we should have no particular reason for expecting a difference and our expected numbers for the two groups would be the same (half the total number of observations each). Thus if our yellow and green groups had referred to pecks by chicks at one of two different coloured grains of rice on a standard background instead of the inheritance of seed colour, our chi-squared table would have looked very different:

	Grain colour	
	Yellow	Green
Number of pecks observed	130	46
Number of pecks expected	88	88

and the result would have been a χ^2 value of 40.09 which exceeds even the threshold value (10.83) for a significance level of 0.1% (0.001) (see Appendix III, Table A). In this case we could safely reject the null hypothesis of no difference in the number of pecks to different coloured grains and infer a bias towards yellow.

An important point to bear in mind with χ^2 is that it is not very reliable when small samples are tested. It is a good idea not to use it when the sample size is smaller than 20 data values or when any expected value is less than 5.

• *Mann–Whitney* U *test.* The Mann–Whitney U test can be performed on raw counts as in the case of chi-squared, except that it deals with each contributing data value in the two groups separately instead of as a single total. Thus if the pecking data in our last example of chi-squared were derived from 10 chicks given the opportunity to peck at yellow grains and 10 given the opportunity to peck at green grains on a standard background, the values that would be used in the chi-squared and U tests might be indicated as follows:

Chick	Pecks to yellow	Chick	Pecks to green	
a	12	k	2	
b	14	l	3	
c	13	m	10	
d	3	n	6	
e	23	o	4	a U test uses
f	13	p	5	these values
g	11	q	3	
h	15	r	1	
i	9	s	7	
j	17	t	5	
Total	130		46	a chi-squared test uses these values (or, in principle, any subtotal of data values)

In addition, however, a U test can cope with both ranking and constant interval measurement data. We can thus use a U test to compare two groups for body size, time spent doing something, percentage responding to drug treatment or any other sort of measurement. Furthermore, unlike some kinds of two-group test working on individual data values, the U test does not require an equal sample size in the two groups. However, the test can be used **only** for comparing **two** groups (we shall return to this point later) and can test **only** a general (or two-tailed) prediction (i.e. there is a difference, but its direction cannot be predicted). It cannot be used when more than two groups are being compared, or to make specific or one-tailed tests. Calculation of the test statistic U is simple and is shown in Box 2:

Box 2

How to do a Mann–Whitney U test

1. Count the number of data values in the group with the fewer values (if there is one); this number is referred to as n_1.
2. Count the number of data values in the other group; this is n_2.
3. Rank all the values **in both groups combined**. The smallest value takes the lowest rank of 1, the next smallest value a rank of 2 and so on. If two or more values are the same they are called **tied values** and each takes the average of the ranks they

Box 2
continued

would otherwise have occupied. Thus, suppose we have allocated ranks 1, 2 and 3 and then come to three identical data values. If these had all been different they would have become ranks 4, 5 and 6. Because they are tied, however, they each take the same average rank of 5 $((4 + 5 + 6)/3)$, though – and this is important – the next highest value still becomes rank 7 just as if the three tied values had been ranked separately. If there had been only two tied values they would each have taken the rank 4.5 $((4 + 5)/2)$ and the next rank up would have been 6. We should thus end up with rank values ranging from 1 to N, where $N = n_1 + n_2$.

4. Add up the rank values within each group giving the total R_1 and R_2 respectively.
5. Calculate $U_1 = n_1 \times n_2 + ((n_1(n_1 + 1))/2) - R_1$.
6. Calculate $U_2 = n_1 \times n_2 - U_1$.
   If U_2 is smaller than U_1 then it is taken as the test statistic U. If not, then U_1 is taken as U.
7. We can now check our value of U against the threshold values in U tables (a sample is given in Appendix III, Table B). If our value is less than the threshold value for a probability of 0.05, we can reject the null hypothesis that there is no difference between the groups. Note that, in this test, we use the two sample sizes, n_1 and n_2, rather than degrees of freedom to determine our threshold value.

If one of the groups has more than 20 data values in it, U cannot be checked against the tables directly. Instead, we must use it to calculate another test statistic, z, and then look this up (some sample threshold values are given in Appendix III, Table C). The calculation is simple:

$$z = \frac{U - \dfrac{n_1 \cdot n_2}{2}}{\dfrac{(n_1)(n_2)(n_1 + n_2 + 1)}{12}}$$

A worked example of a U test can be found in Appendix II.

Tests for a difference between two or more groups

So far, we have introduced significance tests that can test for a difference between two groups. In many cases, of course, we shall be faced with more than two groups. For instance, Fig. 2.1 in the previous section shows the total number of pecks made by chicks towards rice grains of four different colours. If we wanted to know whether there was a significant difference between these colours in the tendency for chicks to peck at them, we should have to deal with four groups of data. How do we do it? The temptation to which many succumb is to do a 'round robin comparison of pairs of groups using a U test or equivalent two-group test. In our Fig. 2.1 example this would mean testing for a difference between red and green grains, then for a difference between green and yellow grains, then for a difference between red and yellow and so on. **The error of this cannot be emphasized too strongly.** The most serious problem arising from such a practice is that it increases the likelihood of obtaining a significant difference by chance when really none exists. To take an extreme example: if we carried out 100 two-group comparisons, then just by chance five of them stand to be significant at the 5% level. Even if we made only 20 comparisons, one is likely to be significant by chance. While this may not seem a serious difficulty when we are dealing with only three or four groups, these examples illustrate the error in principle. To get round the problem, we need tests that can cope with comparisons between several groups at the same time.

• *One-way analysis of variance.* Where we have series of data values falling into a number of groups (in a similar fashion to data values in the two groups of a U test), a one-way analysis of variance (often expressed as the acronym 'one-way ANOVA') is a suitable significance test. There are both parametric analyses of variance and nonparametric tests that have an equivalent function. We shall use the nonparametric versions here partly because of the advantages of nonparametric tests mentioned earlier but partly because it allows specific predictions about the **direction** of differences between groups to be set up in advance and tested (see earlier). This test thus has the great advantage of being usable to test general or specific predictions for two or more groups. For two groups it is therefore more flexible than the U test and as a result we recommend it even in these cases. The test is outlined in Box 3:

How to calculate a nonparametric or -way analysis of variance

1. The test considers i groups of data each of which contains n_i data values.
2. Rank all the values across all groups combined (as in the U test) giving low values low rank scores. Once again (see U test), tied values are given the average of the ranks they would have been ascribed had they been different.
3. Sum the ranks for each group, giving R_i in each case.
4. Make the prediction to be tested. This will be either:
   (a) a general prediction asking if there are any differences at all between the groups, in which case the appropriate test

Box 3
continued

statistic to calculate is H, where

$$H = \frac{12}{N(N+1)} (\Sigma\, R_i^2 / n_i) - 3(N+1)$$

and where N is the total number of data values in all i groups. The significance of the calculated H value can then be looked up as χ^2 for $(i-1)$ degrees of freedom. (Although H is not actually χ^2, its value is distributed in the same way, so it is as if we were using χ^2.);

or

(b) a specific prediction specifying a particular rank order of mean values for the groups. Assign these ranks from the groups with the lowest predicted mean (rank $= 1$) to the group with the highest predicted mean (rank $= i$); these predicted rank values are then the λ_i coefficient values for each group. Using these λ_i values, calculate the following:

$$L = \Sigma\lambda_i R_i$$

$$E = (N+1)(\Sigma n_i \lambda_i)/2$$

$$V = (N+1)[N\Sigma n_i \lambda_i^2 - (\Sigma n_i \lambda_i)^2]/12$$

From L, E and V, the test statistic z can be calculated as

$$z = (L - E)/\sqrt{V}$$

and checked against a table of critical z values (see Appendix III, Table C).

5. Where there are lots of tied values in the data set, it may be necessary to correct for them using a different calculation. However, this only applies where the number of ties is high relative to the total sample size.

A worked example of a one-way analysis of variance for three groups can be found in Appendix II.

• *1 × n chi-squared.* The one-way analysis of variance used a comparison of group means for the rank scores of individual data values to arrive at a test statistic. As in the two-group case, however, we could perform a chi-squared test on the **totals** for each group **as long as the data values are counts (see above)**. We then have what is known as a 1 × n chi-squared analysis (of which the two-group chi-squared is just one form), where n is the number of groups. The calculation of expected values and the test statistic χ^2 is exactly the same as for the two-group chi-squared test; thus we first decide whether expected values should be the average count in each case or whether there are reasons for having unequal expected values, then we calculate $\chi^2 = \Sigma(O - E)^2 / E$ (see Appendix II) and check the outcome in the table (e.g. A in Appendix III) of χ^2 threshold values for $n - 1$ degrees of freedom.

Tests for differences in relation to two levels of grouping

In all the above difference tests, we were concerned with differences within a single level of grouping, e.g. between groupings based on seed colour. However many groups of seed colour we had (red, yellow, green, orange, blue, etc.) we are still dealing only with seed colour and thus with one level of grouping. But we can easily envisage situations in which we would be interested in more than one level of grouping. For example, we might want to know not only whether chicks peck at some colours of grain more than others, but also whether pecking in males is different from that in females. More interestingly still, we might want to know whether the sex of chicks affects the **difference** in pecking at the various colours of grain. Is the difference stronger in one sex? Is it in the same direction in both sexes? And so on. Here we have **two** levels of groupings: seed colour and sex. In the examples that follow, we shall look at analyses that cater for two levels of grouping with two groups in each.

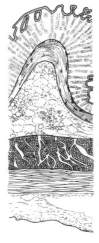

• *2 × 2 chi-squared.* If we have data in the form of counts, we can again use chi-squared, but in a slightly different way from before. The table below shows the number of stomach tissue biopsies revealing cancerous cells in men and women who had been given one of two different chemotherapy treatments:

	Treatment A	Treatment B	Row totals
Men	20 (exp. = 12.3)	17 (exp. = 24.7)	37
Women	0 (exp. = 7.7)	23 (exp. = 15.3)	23
Column totals	20	40	60

Expected values in a 2 × 2 (or any $n \times n$) chi-square table are calculated in a rather different way from those in a $1 \times n$ chi-squared. First the rows and columns are totalled and the grand total calculated. The expected value for each cell in the table is then calculated as (row total × column total)/grand total. Thus for men given Treatment A, the expected number of biopsies revealing cancer is $(37 \times 20)/60 = 12.3$; for women given Treatment B it is $(23 \times 40)/60 = 15.3$. The test statistic is once again calculated as $\chi^2 = \Sigma(O - E)^2/E$, which in this case comes to 18.8. To arrive at the appropriate degrees of freedom, we multiply the number of rows minus one by the number of columns minus one. In a 2 × 2 table, the number of degrees of freedom is therefore $1 \times 1 = 1$; in a 3 × 4 table, it would be $2 \times 3 = 6$ and so on.

Checking $\chi^2 = 18.8$ for l.d.f. against Appendix III, Table A shows that it is significant at the 0.1% (0.001) level. We can thus conclude that men and women differ significantly in their incidence of cancer following the two treatments.

• *2 × 2 two-way analysis of variance.*   As before, if we want to compare sets of individual data values, we can use a nonparametric analysis of variance but this time it is a two-way rather than a one-way analysis. In a 2 × 2 two-way analysis of variance, the data are cast into four cells (two × two groups, which can be cast as two rows and two columns of a table). If we wanted to do a 3 × 5 two-way analysis the data would be cast into 15 cells, and so on for any combination of levels of grouping. Within each cell, it is helpful to have the same number of data values, i.e. the same number of replicates for each combination of groupings. For example, if our two levels of grouping were freshwater fish / marine fish (rows) and male/female (columns), and the variable for comparison was growth rate, it is a good idea to have the same number of measures of growth rate for each combination of water type and sex as shown below (four measures per cell). In nonparametric two-way analyses of variance without equal numbers in each cell, we cannot make general predictions about differences, but only specific ones. While we are emphasizing the value of making specific predictions here, it is, of course, often useful (or even necessary) to be able to make general predictions too.

| | | Water type | |
		Fresh	Marine
	Male	a b c d mean = A	e f g h mean = B
Sex			
	Female	i j k l mean = C	m n o p mean = D

As in the one-way analysis of variance, predictions can be general or specific with different test statistics being used in each case. Now, however, general predictions are concerned with the following: (a) differences in column means (i.e. differences within one level of grouping, say between freshwater and marine environments); (b) differences in row means (i.e. differences within the other level of grouping, in this case sex); and (c) any interaction between the levels of grouping (e.g. is the effect of sex greater in one environment than in the other?). Making specific predictions is rather more involved because we need to be clear about exactly what we are comparing and to calculate different coefficients for each specific comparison so that they can be tested. In the one-way analysis of variance, calculating the coefficients for testing was easy because they were simply the predicted rank order of the mean values for the groups being compared (see Box 3). In a 2×2 comparison, there are four cells thus allowing three classes of specific prediction (generally, if i is the total number of groups then $i - 1$ classes of specific prediction can be made).

How to calculate a nonparametric 'two-way analysis of variance'

Testing general predictions:

1. Rank the data values in all cells combined and add up the ranks in each cell to give a rank total R for that cell.
2. We can now use these rank totals to test our **general** predictions:

   a) *Marine fish differ from freshwater fish.*
      Sum the rank totals for each **column** (see table above) separately giving an R_i value for marine and an R_i value for freshwater environments. Now calculate H as in the one-way analysis of variance (Box 3), using the column R_i values and their appropriate n_i values (here there are 8 values making up each R_i, hence $n_i = 8$), and check the resulting value in the appropriate table as χ^2 for $(i - 1)$ degrees of freedom.
   (b) *Male fish differ from female fish.*
      Sum the rank totals for each **row** separately giving an R_i value for male and an R_i value for female fish and calculate H again as above. Again, $n_i = 8$ in our example.
   (c) *There is an interactive effect of water type and sex on the growth rate of fish.*
      This is an open-ended general prediction which combines both (a) and (b) above. It is asking whether there is any interaction between levels of grouping in determining growth rate. The calculation of H is exactly as above but

Box 4
continued

the $\Sigma R_i^2/n_i$ term includes the rank totals for **all** the cells in the table instead of just the columns or the rows to give H_{tot}. n_i is again the number of values making up each R_i ($=4$ in our example). H for the interaction, H_{int}, can then be calculated as: $H_{int} = H_{tot} - H_{water} - H_{sex}$, where H_{water} and H_{sex} refer to H values from (a) and (b) above. The degrees of freedom for H_{int} are $df_{tot} - df_{water} - df_{sex}$, in this case, therefore, $3 - 1 - 1 = 1$ (df_{tot} is the total i (number of group means) minus 1).

Testing specific predictions:

3. Testing our **specific** predictions is a little more complicated but still relatively straightforward. We give three illustrations below:

The three classes of specific prediction are:

    (a) A prediction about the columns, e.g. *Marine fish grow faster than freshwater fish* (or *vice versa*). In terms of the means in the table above this predicts that $(B+D)>(A+C)$. Another possibility is the opposite prediction, that freshwater fish grow faster than marine, i.e. $(A+C)>(B+D)$.

    (b) A prediction about the rows, e.g. *Male fish grow faster than female fish* (or *vice versa*). This predicts $(A+B)>(C+D)$ (or, conversely, for the prediction that females grow faster than males, $(C+D)>(A+B)$).

    (c) A prediction about interaction, e.g. *The effect of water type on growth rate will be greater in male fish than in female fish.* This predicts that $(A - B) > (C - D)$. The converse (the effect will be greater in females) would, of course, predict $(C-D)>(A-B)$. This class of prediction is thus concerned with the *interaction* between water type and sex.

These are the predictions we can make about the relative sizes of the means; how do we arrive at the coefficients for testing? The procedure is as follows:

The first step is to rearrange the various predicted inequalities so that all the means are on the left, thus:

(a) The prediction about the effect of water type becomes $-A + B - C + D > 0$ (for growth in marine > growth in freshwater) or $+A - B + C - D > 0$ (for growth in freshwater > growth in marine).

    We then substitute 1 with the appropriate sign for each letter so that we arrive at $-1, +1, -1, +1$ or $+1, -1, +1, -1$ respectively.

(b) In the same way, the prediction about the effect of sex becomes $+A + B - C - D > 0$ (for males > females) or $-A - B + C + D > 0$ (for females > males) and the coefficients thus $+1, +1, -1, -1$ or $-1, -1, +1, +1$.

Box 4
continued

(c) The interaction predictions must also be framed in this way. Thus the prediction $(A - B) > (C - D)$ becomes $+A - B - C + D > 0$ and the coefficients $+1, -1, -1, +1$. The prediction $(C - D) > (A - B)$ becomes $-A + B + C - D$ and the coefficients therefore $-1, +1, +1, -1$.

Then, testing these predictions

(a) *Marine fish grow faster than freshwater fish.*
Remember this means we are testing the prediction $-A + B - C + D > 0$. The coefficients λ_i thus become $-1, +1, -1, +1$ so that the rank totals for each cell are weighted as follows:

$$L = \Sigma \lambda_i R_i$$

$$= (-1)(R_{freshwater/male}) + (+1)(R_{marine/male})$$

$$+ (-1)(R_{freshwater/female}) + (+1)(R_{marine/female})$$

E and V can then be calculated as

$$E = (N + 1)(\Sigma n_i \lambda_i)/2$$

$$V = (N + 1)[N\Sigma n_i \lambda_i^2 - (\Sigma n_i \lambda_i)^2]/12$$

The test statistic z can then be calculated as

$$z = (L - E)/\sqrt{V}$$

and checked against a table of z values (see Appendix III, Table C).

(b) *Male fish grow faster than female fish.*
Now we are testing the prediction $+A + B - C - D > 0$, so λ_i becomes $+1, +1, -1, -1$. The calculation of L, E and V and then the test statistic z can proceed as above, but with the new λ_i weightings.

(c) *The effect of water type is greater in males than in females.*
This tests the interaction prediction $+A - B - C + D > 0$ using λ_i of $+1, -1, -1, +1$. Once again, follow the calculations above for L, E, V and z.

A full worked example of a 2×2 two-way analysis of variance can be found in Appendix II.

Tests for a trend.

As with analysis of differences, there are many tests that cater for trends. We shall introduce two simple ones here, both looking at the relationship between two sets of data. More complex versions of these kinds of test allow multiple relationships to be tested at the same time, but they are beyond the scope of this book.

• *Spearman rank correlation.* The first test is a nonparametric test of correlation. As its test statistic, it calculates the Spearman rank correlation coefficient r_s. A correlation coefficient (there are a number of different ones) quantifies the extent to which two sets of values vary together. A large **positive** coefficient indicates a strong tendency for high values in one set to co-occur with high values in the other and low values in one set to co-occur with low values in the other. A large **negative** coefficient indicates a strong tendency for high values in one set to co-occur with low values in the other and vice versa. Correlation coefficients take a value between $+1.0$ and -1.0, with values of $+1$ and -1 indicating respectively a perfect positive or negative association. 'Perfect association' means that every value in one set is predicted perfectly by values in the other set. In a rank order correlation, like the Spearman, this means that where $r_s = +1.0$ the rank orders of the two sets of data values are the same, i.e. rank 1 in set 1 corresponds with rank 1 in set 2 and so on through all the other ranks. A coefficient of 0 indicates there is no association between the two sets of values so that values in one set cannot be predicted by those in the other. If a correlation is significant, it implies that the size of the coefficient differs significantly (positively or negatively) from zero, the value expected under the null hypothesis.

The Spearman rank correlation is one of the two most commonly used tests of correlation, the other being the parametric Pearson (or product moment) correlation which yields the coefficient r. Being a nonparametric test, the Spearman rank correlation can work with ordinal (ranking) or constant interval measurements and, of course, is not sensitive to departures from normality in the distribution of data values. Calculating the coefficient is straightforward (Box 5).

How to calculate a Spearman rank correlation coefficient   **Box 5**

1. Set out the two sets of data values to be correlated in pairs (remember, for each value in set 1 there must be a corresponding value in set 2). Thus, if we were looking for a correlation between height and weight in people, the data would be set out as below:

Person	Weight (kg)	Height (m)
1	63	1.8 (pair 1)
2	74	1.7 (pair 2)
3	60	1.9 (pair 3)
4	71	1.8 (pair 4)
.	.	. .
.	.	. .
.	.	. .
		etc.

Box 5
continued

It may be that several values of one measure are paired with the same value of the other, for example when measuring some behaviour in several individuals from the same social group and using these values in a correlation of time spent doing the behaviour and group size. In this case, the data might be as follows:

Observation	Time spent in behaviour (s)	Group size
1	15.3	3 (pair 1)
2	17.1	3 (pair 2)
3	18.0	5 (pair 3)
4	6.0	5 (pair 4)
5	31.1	5 (pair 5)
.	.	. .
.	.	. .
.	.	. .
		etc.

2. Rank the values for the first measure only, then rank the values for the second measure only.
3. Subtract second-measure ranks from first-measure ranks (giving d_i) then square the resulting differences (d_i^2) and calculate the Spearman coefficient as:

$$r_s = 1 - [(6\Sigma_i d_i^2)/(n^3 - n)]$$

where n is the number of pairs of data values.
4. If n is between 4 and 20 (4 is a minimum requirement), consult Appendix III, Table D for the appropriate sample size to see whether the calculated r_s value is significant. Remember that general (two-tailed) or specific (one-tailed) tests have different threshold values for r_s. Thus you must know whether you are looking for any association at all (general), or just a positive, or just a negative one (specific).

5. If n is greater than 20, a different test statistic, t, is calculated from r_s as:

$$t = r_s \sqrt{[(n - 2)(1 - r_s^2)]}$$

t can then be checked against its own threshold values (Appendix III, Table E) for $n - 2$ degrees of freedom.

A significant calculated value for r_s or, with large samples, t, allows us to reject the null hypothesis of no trend in the relationship between our two measures. A fully worked example of a Spearman rank correlation analysis is given in Appendix II.

While we can test for a trend with correlation analyses, we must interpret them with care. Two things in particular should always be borne in mind. First, a correlation does not imply cause and effect. While we may have reasons for **supposing** that one measure influences the other rather than vice versa (see earlier discussion of x and y measures in trend analysis), a significant correlation cannot be used to confirm this. All it can do is demonstrate that two measures are associated. A well-known example illustrates the point. Suppose we acquired some data on the number of pairs of storks breeding in Denmark each year since 1900, and also the number of children born per family in Denmark in the same years. Plotting a scattergram, with breeding storks as the x-axis and babies as the y-axis, and calculating a Spearman correlation coefficient reveals a significant positive correlation at $p < 0.001$ between the two measures. Do we conclude that storks bring babies? Of course not! All we can conclude is that, over the period examined, there is some association between the number of breeding storks and the human birth rate; perhaps both species simply reproduce more during long, hot summers! Second, correlation analyses assume that associations between measures are linear (or reasonably so) if the test is parametric, or at least monotonic (continuously increasing or decreasing) if the test is nonparametric. If they are not, a lack of significant correlation cannot be taken to imply a lack of association. This is made clear in Fig. 3.4 in which the relationship between two measures is U-shaped. A correlation coefficient for this would be close to zero, but this doesn't mean there is no association. The book by Martin and Bateson (1986) contains a very useful discussion of these and other problems concerning correlation analyses.

• *Linear regression.*   Finally we come to the only parametric significance test in the book – linear regression. The Spearman rank correlation, like other correlation analyses, allows us to judge whether two measures are associated, but not much more. In particular, it does not allow us to establish whether changes in the value of one measure **cause** changes in the value of the other.

In the case of the storks and babies above, it is fairly clear that changes in x (the number of breeding storks) do not cause changes in y (the number of human babies): the association arises either through some indirect cause-and-effect relationship (e.g. hot, sunny years encourage both storks and humans to breed) or through trivial coincidence. In many cases, though, it is not so clear whether there is or is not a direct cause-and-effect relationship between x and y. The best way to decide is to do an experiment where, instead of merely measuring x and y as in correlation analysis, we experimentally change x-values and measure what

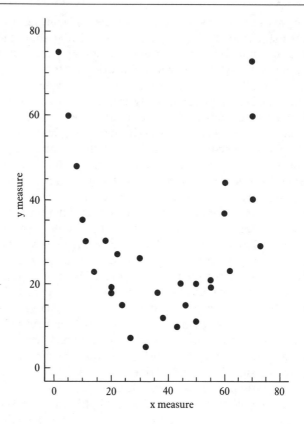

Fig. 3.4

subsequently happens to y. If we see that y changes when x is changed, we can be reasonably happy with a cause-and-effect interpretation. This is quite different from correlation analysis. In addition, correlation analysis does not allow us to say much about the quantitative relationship between x and y (if x changes by n units, by how much does y change?) or allow us to predict values not included in the original trend analysis (e.g. can we predict the response of an insect pest to a 40% concentration of pesticide when an analysis of the effect of pesticide concentration goes up to only 30%?). With certain qualifications, **linear regression analysis** may allow us to do all the above. The qualifications arise mainly from the fact that regression is a parametric test; it thus demands constant interval measurement and a reasonable approximation to normality in one of the data sets (the y measure, see below) it is relating. Like parametric correlation analysis, and as its name implies, linear regression also requires the relationship to be passably linear, though there are ways of overcoming some non-linearity, for instance by log-transforming data sets.

Regression analysis proceeds as follows: the x values are decided upon in advance by the investigator to cover a range of particular interest, and y values are measured in relation to them to see how well they are predicted by x. Because x values are selected by the investigator, they are unlikely to be normally distributed, hence the requirement of normality only on the y measure. Linear regression then calculates the position of a line of best fit through the data points and uses the equation for this line (the **regression equation**) to predict other values for testing. The criterion of 'best fit' in this case is the line that minimizes the magnitude of positive and negative deviations from it in the data (a more precise, mathematical way of doing what we attempt to do when drawing a straight line through a scattergram by eye). The significance of a trend can be assessed in one of two ways: the first is based on the difference in the **slope** of the best fit line from zero (a line with zero slope would be horizontal) and is indicated by the test statistic t; the second tests whether a significant amount of variation in y is accounted for by changes in x, and is indicated by the test statistic F. Generally speaking, if the trend fails to reach significance, the best-fit line should not be drawn through the points of the scattergram.

We shall use F as our test for a linear regression. The procedure for calculating F looks involved but is in fact very straightforward. The essential steps are outlined in Box 6. Reference to some of the jargon of parametric statistics is unfortunately inevitable, but can largely be ignored for our purposes. Its inclusion, however, may aid any further reading around:

Summary of linear regression analysis   **Box 6**

1. Calculate the square of every x and y value to give x^2 and y^2.
2. Calculate the product xy for every pair of x and y data values.
3. Calculate a value known as the **sum of squares of x**, S_{xx}, as: $\sum x^2 - (\sum x)^2/n$, where n is the number of pairs of data values.
4. Calculate the **sum of squares** of y, S_{yy}: $\sum y^2 - (\sum y)^2/n$.
5. Calculate the **sum of the cross-product**, S_{xy}: $\sum xy - (\sum x)(\sum y)/n$.
6. Calculate the slope of the line as: $b = S_{xy}/S_{xx}$.
7. Calculate the intercept of the line on the y-axis as: $a = \bar{y} - b\bar{x}$, where $\bar{y}$ is the mean of the y values and $\bar{x}$ is the mean of the x values.

The line can now be fitted by calculating $y = a + bx$ for some sample x values and drawing it on the scattergram.

Box 6
continued

To calculate the standard error of the slope:

8. Calculate the **variance of y for any given value of x** as:
    $s^2_{y/x} = [1/(n-2)][S_{yy} - S^2_{xy}/S_{xx}]$.
9. The **standard error of the slope** is then: $\sqrt{(s^2_{y/x}/S_{xx})}$.
10. To find the test statistic F, calculate the following:

    Regression sum of squares (RSS) $= (S_{xy})^2/S_{xx}$
    Deviation sum of squares (DSS) $= S_{yy} - (S_{xy})^2/S_{xx}$
    Regression mean square (RMS) $=$ RSS$/1$

    (1 is the value always taken by the **regression degrees of freedom**.)

    Deviation mean square (DMS) $=$ DSS$/(n-2)$.

    ($n-2$ is the value taken by the **deviation degrees of freedom**.)

F is now calculated simply as $F =$ RMS/DMS and its value can be checked against critical values in F-tables (Appendix III, Table F) for 1 (f_1) and $n-2$ (f_2) degrees of freedom.

11. To find y for new values of x:
    Having established our regression equation, we might well want to predict y for other values of x that lie within the range we actually used in the analysis. Once we had then gone away and **measured** y for our new x-value we should want to see whether it departed significantly from its predicted value. Three steps are needed:
    (a) calculate the predicted y-value using the equation $y = a + bx$ as when fitting the regression line, but this time use the new x-value (x') in which you are interested;
    (b) calculate the standard error (s.e.) to the predicted y-value as follows:

    $$\text{s.e.} = \sqrt{[(s^2_{y/x})(1 + 1/n + d^2/S_{xx})]}$$

    where $d = x' - \bar{x}$;
    (c) calculate the test statistic t as:

    $$t = \frac{\text{observed } y - \text{predicted } y}{\text{s.e.}}$$

    and look up the calculated value of t in t-tables (Appendix III, Table E) for $n-2$ degrees of freedom (where n is the number of pairs of data values in the regression). If t is significant it means the measured value of y departs significantly from the value predicted by the regression equation and might lead to interesting questions as to why.

A worked example of a regression analysis can be found in Appendix II.

TESTING HYPOTHESES

The previous section has introduced an armoury of basic significance tests with which we can undertake confirmatory analyses of differences and trends. Knowing that such tests are available, however, is not much use unless we know how and when to employ them and gear our data collection to meet their requirements. It is important to stress again, therefore, that the desired test(s) should be borne in mind from the outset when experiments and observations are being designed and the data to be collected decided upon.

Deciding what to do

Having arrived at some predictions from our hypotheses, we must decide how best to test them. This sounds straightforward in principle but involves making a lot of careful decisions. Are we looking for a difference or a trend? What are we going to measure? How are we going to measure it? How many replicates do we need? What do we need to control for? There is no general solution to any of these problems; the right decision depends entirely on the prediction in hand and the material available to test it. In a moment, we shall go back to the predictions we derived from our observational notes to see how we can test some of these. Before doing that, however, we should be aware of some important pitfalls of experimental/observational design and analysis.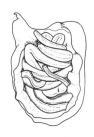

Some dangerous traps

• *Confounding effects.* One of the commonest problems in collecting and analysing data is avoiding so-called confounding effects. Confounding effects arise when a factor of interest in an investigation is closely correlated with some other factor which may not be of interest. If such a correlation is not controlled for, either in the initial design of an investigation or by using suitable techniques during analysis, the results will inevitably be equivocal and any potential conclusions compromised. For example, suppose we wanted to know whether the burden of a particular parasitic nematode increased with the body size of the host (e.g. a mouse). We might be tempted simply to assay worm burden and measure body size for a number of arbitrarily chosen host individuals, and then perform a correlation analysis. If we got a significant positive correlation coefficient, we might conclude that burden increased with body size. An unwelcome possibility, however, is that host body size correlates with age so that bigger hosts also tend to be older. If there is some age-related change in immune competence (e.g. older hosts are less able to resist infection), a positive correlation with size could arise that actually has nothing to do with host body size. In this case size is confounded with age and simple correlation analysis cannot disengage the two. The best solution here would be to select different sized hosts from a given age group so that the confounding effect is controlled for from the outset.

• *Floor and ceiling effects.* Floor and ceiling effects arise when observational or experimental procedures are either too exacting or too undemanding in some way to allow a desired discrimination to be made. For example, looking for differences in mathematical ability among people by asking them the solution to $5 + 3$ is unlikely to be very fruitful because the problem is too easy; everyone will get the right answer straightaway. A ceiling effect (everyone performs to a high standard) will thus prevent any differences there may be in ability becoming apparent. Conversely, if the same people were asked to solve a problem in catastrophe theory the odds are that no-one would be able to do it. In this case, a floor effect (everyone does badly) is likely to prevent discrimination in ability. Floor and ceiling effects are not limited to performance-related tasks. Similar limitations could arise in, for instance, histological staining. If a particular tissue requires just the right amount of staining to become discriminable from other tissues, the application of too little stain would result in everything appearing similarly pale (a floor effect) while too much stain would result in everything appearing similarly dark (a ceiling effect). Real differences in tissue type would thus not show up at the extremes of stain application. Floor and ceiling effects are clearly a hazard to be avoided and are well worth testing for in preliminary investigations.

• *Non-independence.* One of the commonest sources of error in data collection and analysis arises from non-independence of data values. In many circumstances, there is a temptation to treat repeated measures taken from the same subject material as independent values during statistical analysis. As Martin and Bateson (1986) point out, this error arises from the misconception that the aim of a scientific observation or experiment is to obtain large numbers of measurements rather than measurements from a large number of subjects. The point is, of course, that obtaining additional measures from the same subject is not the same as increasing the number of subjects in the sample. An example of such an error would be as follows. Suppose an investigator wished to assess the average rate of nutrient flow in the phloem of a particular plant species. Setting up a preparation might be involved and time-consuming. To save effort, the investigator decides to take as many measurements as possible from each preparation before discarding it. As a result, there are 15 measurements from one preparation, 10 from another and 12, 16 and 5 from three more. To calculate the average, the investigator totals the measurements and divides by $n = 58$ (i.e. $15 + 10 + 12 + 16 + 5$). Of course, the measurements from each preparation are not independent; there may be something about the plant in each case that gives it an unusually high or low rate of nutrient flow relative to most of the plants in the population. Incorporating each measurement taken from it as an independent example of flow rate in the population as a whole is clearly going to bias the average upwards or downwards. The true n size in the above example is 5 (the number of preparations) not 58. Measurements from each preparation should thus be averaged, or collapsed in some other way, to provide a single value for use in analysis. The fallacy of this kind of pseudoreplication becomes obvious if we consider estimates of average plant height rather than nutrient flow rate. Few people would seriously measure the height of the same plant 16 times and regard these as independent samples of the height of the species concerned. The principle, however, is exactly the same in the flow rate example and is referred to as pseudoreplication.

The problem with non-independence is that it can operate at several different levels and sneak insidiously into analyses unless careful attempts are made to exclude it. We have discussed it only at the level of the individual subject. However, depending on what is being measured, it could arise if, for example, related individuals are used as independent subjects or if plants grown on the same seed tray or animals kept in the same cage are used. We should thus be on our guard against it at all times.

TESTING PREDICTIONS

Having highlighted some potential pitfalls, we must now bear them in mind as we return to our main observational examples and design experiments to test some of their predictions. We shall take one prediction from each.

E.g. Prediction 1A(ii) *Males will display more vigorously to larger females.*

Example 1

Guppies

This predicts a positive trend between male display vigour and female size. It invites us to pick a number of different sized females and measure

the vigour with which males display towards them. Since we are going to select females by size, and it would be useful to have a quantitative measure of any relationship between female size and male display so we can test its predictive value, regression analysis seems to be the obvious test to go for. The questions that then arise are: (a) How do we quantify female size? and (b) How do we measure male display vigour?

Quantifying female size. We could measure female size in several ways: snout–tail length, weight, volume, dorso-ventral depth at the widest point – which should we use? A good idea is to think back to the reason for making the prediction in the first place. The expectation that males would display more to bigger females was based on the assumption that the size of a female was related to the number of eggs she was carrying. Since carrying lots of eggs increases a female's bulk, it would seem sensible for bulk to be expressed somehow in our measure. One choice might be the volume of the female (e.g. measured by water displacement in a graduated beaker or cylinder) since this should increase with the number of eggs. However, a simple measure of volume could lead to large females that were not carrying eggs getting the same score as smaller females that were. A better measure would be one that controlled for this confounding effect of body size. One possibility is the **ratio** of volume (V) to length (L), i.e. (V/L); on this basis, a given volume would be devalued by increasing length. Of course, this is still taking a shot in the dark. Unless we know something about how different measures of size **actually** vary with number of eggs, any measure we adopt is likely to have a hefty margin of error. Our ratio measure, for instance, could undervalue a large female with exactly the same number of eggs as a smaller (but higher ratio) female. From a male's point of view both females might be equally desirable because both offer the same potential number of offspring. Ideally, then, we ought first to investigate the relationship between measures of size and number of eggs (see Basic Considerations in Chapter 1).

Measuring the vigour of male display. Deciding on a measure of male display vigour may be even trickier. Potentially at least, there are innumerable ways it could be measured – the number of visits to a female, the time spent displaying, the number of times a display is repeated on a visit, the number of display components performed, the percentage of visit time spent displaying, to mention a few. Since it is variation in the level of interest shown towards different females that matters here, however, we should pick a measure that seems a credible yardstick of sexual interest rather than something that could just as easily reflect non-sexual social interest or simply curiosity about a novel object appearing in the environment. On this basis, therefore, it seems better to go for measures of display performance (e.g. time spent in display, number of repeats of display, number of display components performed) instead of simply measures of the tendency to associate (e.g. number of visits to the female, time spent near the female). However, once again, we might want to play about a bit more with interactions between males and females to see what could best be measured before deciding.

Designing the procedure.   Having decided on our measures, we now have to decide on the experimental procedure. How are males and females to be introduced? How many males should be presented to each female? How long should interactions be allowed to go on?

Clearly, just introducing a mixed group of males and females to an aquarium and measuring interactions as they happen to occur is unlikely to get us far because both male and female responses may be compromised by interference. It would be better to introduce fish in a more controlled way, but how? Again, we should go back to our prediction. The prediction concerns the response of males to female size. If we let fish actually meet and interact physically there is a danger that the response of the male will be influenced by that of the female: for example, if the female approaches the male rather than swimming away, she may encourage more vigorous display than the male would have performed on the basis of her size alone. This potential complication could arise even if females were placed on the other side of a clear partition as in the initial observations. One way round the difficulty might be to confine the female in a beaker or jar within the aquarium so she is limited in the movements she can make when the male approaches.

Another unwelcome influence on the behaviour of the fish might be the disturbance and stress of being introduced to the test aquarium. Some kind of settling period, say 5 minutes, would thus seem to be a good idea. However, we shouldn't want test males and females to be aware of each other while they were settling down so it would be useful to separate them with an opaque partition that can be removed at the start of observations.

The next problem concerns how many males and females to use and whether to present each male to each female or use unique pairings. Clearly we want a good range of female sizes that includes some extremes as well as sizes around the average. The range we choose will obviously be constrained by what is available, but somewhere between six and ten different sizes seems sensible. This will then determine the number of points we shall have along the x-axis of our scattergram. We could present a single unique male to each female, but this would give us only the same number of y values as x values (i.e. 6–10) and would confound female size with individual male (an apparent trend might arise because some of the larger females, say, just happened to be tested with unusually vigorous males). Presenting the same males, perhaps five, to **each** female would control for chance individual effects but brings with it a new problem – the order in which to present them to different females. A naive experimenter might be tempted to present males to females in order of increasing female size, i.e. each male would encounter the smallest female first, then the next one up and so on until the largest. The obvious shortcoming with this is that it confounds female size with a male's experience of the test procedure. Males may become more used to being manipulated with successive tests and end up displaying more vigorously in later tests because they are less disturbed. Conversely, they may gradually learn that, while the female is visible, she is actually unobtainable, or will be unresponsive because she is confined. Males might thus become **less** vigorous in their displays later in the test sequence. We can overcome these potential order effect problems by presenting different sized females to each male in a scrambled, arbitrary order that differs between males. (People often refer to this as 'randomizing' treatments. However, 'random' has a very specific, technical meaning within statistical theory and what are often called random distributions or random orders are usually no such thing. Unless distributions or orders have been produced using random number generators or tables, they are unlikely to be truly random. People thus often use 'random' when they really mean 'arbitrary', a haphazard disordering to reduce systematic pattern.) We might also want to consider changing the water in the aquarium between tests to avoid any effects of chemical cues emanating from previous males.

Finally, we must decide how long interactions should last. Since we are after the male's response to female size, we again want to avoid any complicating effects of the females' own responses (or lack of them). Consequently, it might be best to record a male's displays over a relatively short period, say 2–3 min, after he first approaches the female. As we are interested in the fine detail of male display (number of displays, number of different display components etc., see above) during the observation period, dictating the sequence of male behaviour into a tape recorder and transcribing the tape later to obtain various measures of display intensity would reduce the risk of missing things.

Testing our simple prediction has involved us in some detailed considerations of measurement and experimental design. Such considerations are not unique to the guppy example or to the study of behaviour; they apply in principle to any investigation we might undertake. Without this kind of careful thought, it is clear that we risk making blunders that could seriously compromise or negate any useful conclusions. The same kind of problems, though different in detail, arise in our other examples.

E.g. Prediction 2B *Chicks will find food faster if they forage with companions.*

Example 2

Chicks

This predicts a difference in the speed with which chicks find food when searching alone as opposed to searching in the presence of other chicks. Since we are likely to be dealing with some kind of constant interval measurement (speed of finding food) and not counts, a U test or one-way analysis of variance for two groups are the obvious significance tests to choose. How, then, are we going to measure speed of finding food and arrange solitary versus group foraging?

Although the prediction is derived from a hypothesis about the potential foraging information provided by companions, it predicts only a very general consequence of receiving such information: a subject chick will find more food over a given period of foraging. We could thus test it simply by allowing a group of chicks to search for food items (grain scattered in an arena, as in the initial observations) and comparing the feeding rate of these chicks with that of solitary birds in the same apparatus. We shall see later how the hypothesis and its predictions can be refined.

Since our initial observations suggested that grains which were a similar colour to the arena background were harder to find, it might be worth using these to avoid a possible ceiling effect; if grains were very easy to see, additional information from other foragers wouldn't be much use and, all other things being equal, solitary chicks might be expected to do just as well as those in a group. Using similar-coloured 'cryptic' grain might thus increase the chance of revealing an effect of companions on feeding rate.

How many chicks should be used in the group treatment and how long should subjects be observed foraging? The number of chicks per group should be sufficient to generate information about food but not so large as to eat all the food too quickly for subjects to benefit or cause interference. In an arena of the size used in the initial observations, a group of four would probably be best. Just as decreasing food availability with time influences the number of chicks we choose, it also influences the length of time for which we observe chicks and the number within groups that are chosen as subjects. Observation times should be long enough to allow subjects to pick up and act on information from companions but there is little point in making them so long that most of the food has gone and/or chicks are no longer looking for food. Some trial-and-error observations might be a good idea here. For the same reason, and to maintain independence between replicates of group observations, it would make sense to use one arbitrarily chosen chick as a subject. These, and the chicks used as solitaries, could then be observed for the chosen amount of time and the number of rice grains pecked by them noted. While it might seem a good idea to use the same chick as both solitary and companion subjects, there may be an effect of experience with the foraging task and/or satiation on whichever treatment came second. A systematic effect could be controlled for by scrambling the order of solitary and group treatments between subjects, but the problem could be avoided altogether by arbitrarily allocating different chicks to the solitary and companion treatments at the outset. For similar reasons, different companion chicks should also be used in each companion treatment. Somewhere in the region of 8–12 replicates of solitary and companion treatments would be reasonable for a U test or two-group one-way analysis of variance (though, of course, neither test demands the same number of replicates in the two groups).

Example 3

Sticklebacks

E.g. Prediction 3A *Time spent motionless by individuals will decrease with increasing numbers of fish but will be greater for any given number if fish are alarmed.*

This can be interpreted as predicting both a trend (group size effect) and a difference ('not alarmed' versus 'alarmed'). The trend would be testable by either a Spearman rank correlation or regression, so we could use the latter and gain the extra quantitative information it gives us (as long as our data meet its requirements, see earlier). The difference could be tested using a U test or two-group one-way analysis of variance on replicates of a given number of fish that had either been alarmed in some way or not alarmed. However, this is a rather clumsy way of handling the analysis. Instead of doing a regression and then, say, a U test, we could analyse both effects in one go by using a two-way analysis of variance. Our two levels of grouping in this case would be 'number of fish' and 'alarmed/not alarmed'. If we opt for the two-way analysis of variance, we should be using different numbers of fish as groups in a difference analysis rather than as points along the x-axis of a trend. Thus, we could choose, say, three different numbers of fish, e.g. 4 (group 1), 8 (group 2) and 12 (group 3). Obviously, we should lose the trend information yielded

by the regression, but we should still discover whether there is a significant effect of group size. We should also discover whether there is (a) an effect of alarm that is independent of the number of fish and (b) some interaction between alarm and number of fish (e.g. does the effect of alarm decrease with increasing numbers of fish?). If we are to make any **general** predictions from our two-way analysis of variance, however, we must have the same number of data values in each of the cells (see earlier). For a 3-group (number of fish) × 2 group (alarmed/not alarmed) analysis, that means the same number in each of six cells. If we decide on four replicates per cell, we should need four values for four fish not being alarmed, four for four fish being alarmed and the same for the two eight-fish and 12-fish cells. Moreover, the fish used to generate each replicate value in a cell ought to be different so that replicates are truly independent.

Having decided on the analysis, we must now decide how to measure periods of motionlessness. The first thing to consider is when, after introducing fish to the test aquarium, we should take measurements. Some kind of settling period seems advisable, though sudden transference to the aquarium could itself be used to induce alarm and thus constitute the 'alarmed' treatment. After allowing fish to settle, we could measure time spent motionless in a number of ways. One way would be to select arbitrarily a focal fish and record the total amount of time it spends motionless during a given period of observation; a single fish would thus provide the data value for each replicate within the cells of the analysis of variance. If only one fish is used, however, it would be wise to increase the number of replicates within cells to allow for chance individual variation. A way round this inconvenience would be to time motionlessness in a given number of arbitrarily chosen fish in each group (say three) and use the mean or median time across fish as the data value for each replicate.

To investigate the effects of alarm, it would be best to use separate groups of fish for each treatment. If we used the same group in both 'non-alarmed' and 'alarmed' treatments, we should have to vary the order of treatments across groups to avoid confounding treatment with the amount of time fish will have spent together. However, this would mean some fish being alarmed first and thus possibly associating the test aquarium environment with alarm in their second, 'non-alarmed', treatment. Such a carry-over effect would reduce the likelihood of detecting a real difference between 'non-alarmed' and 'alarmed' treatments. What kind of alarm stimulus should we use? Although transference to the test aquarium could be used (see earlier), and the behaviour of 'alarmed' fish measured immediately after being introduced, this may not be the best option for two reasons: (a) 'alarm' is confounded with the novelty of the social and physical environment which might itself affect behaviour (e.g. fish may show more exploration early on and thus spend less time motionless) and (b) sudden transference to a new environment is not something that is likely to be experienced by fish naturally, so they may not have evolved any particular response to it (see Basic Considerations in Chapter 1). A better idea might thus be to present a naturalistic alarm stimulus after fish have had a chance to settle. A little reading around will reveal that sticklebacks are preyed upon by a number of fish and bird predators. Mimicking an attack by one of these could be useful. Perhaps the simplest thing would be to pass a hand or other object over the aquarium to create a sudden shadow like that of an attacking bird. The behaviour of fish could then be recorded immediately after a pass.

Example 4

Crickets

E.g. Prediction 4C *Encounters will progress further when opponents are more similar in size and it is more difficult to judge which will win.*

This prediction derived from observing that encounters between male crickets followed an apparently escalating pattern from chirping and antenna-tapping to out-and-out fighting and that, on the whole, bigger crickets tended to win. A possibility, therefore, is that progressive escalation reflects information-gathering about the relative size of an opponent and the likelihood of winning if the encounter is continued. If relative size is difficult to judge, as when two opponents are closely matched, the likelihood of winning cannot be judged in advance and the only way to decide the outcome is to fight. The prediction is thus of a negative trend between degree of escalation and the relative size of opponents: degree of escalation should increase with decreasing difference in size.

At first sight, this seems easy enough to test using a Spearman rank correlation or regression analysis. However, we first need some way of measuring degree of escalation. So far all we have are behavioural descriptions – chirping, antennating, fighting, etc. – from which we have inferred levels of escalation. Somehow we must put numbers to these. It is clear that we cannot put the behaviours on some common constant interval scale; we cannot, for instance, say that antennating is twice as escalated as chirping and fighting ten times as escalated. The easiest thing

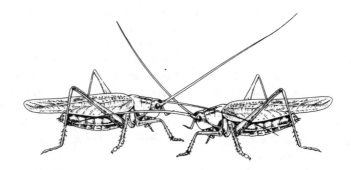

is simply to rank them. Thus what we assume to be the lowest level of escalation, say chirping, takes a rank of 1 and the highest level a rank of n, where n is the number of levels we decide to identify. We can then use the ranks of 1 to n as the y values in our trend analysis.

To obtain our x values, we must decide on a suitable measure of size. Ideally, the measure should be reliable and repeatable within and between individuals; measuring the size of the flexible abdomen, for instance, might not be a good idea because this could vary with food and water intake and thus vary from one encounter to another. It would be better to measure some component of the hard exoskeleton, e.g. the length of the long hind leg or the width of the thorax, which will not vary over the time course of observations. Of course, in our analysis, we are interested in a measure of the **relative** size of opponents, so our x values must be some measure of relative size. The most obvious might be the **difference** in size between opponents. However, it is not hard to see why this would be inadequate. Suppose we observed two crickets of thorax widths 7.5 and 8.5 mm respectively. Suppose we observed another pair of thorax widths 6.5 and 5.5 mm. In both cases, opponents differ by 1 mm and would score the same on a simple difference measure. In the first case, however, 1 mm is only 6% of the combined width measures; in the second it is 8%. A 1 mm difference may thus create a greater asymmetry in the likelihood of winning in the second case than in the first. As a result it would be better to use a ratio rather than a difference scale on the x-axis, e.g. size of bigger opponent/size of smaller opponent.

Having decided on our measures, we can now plan observations. Since we are looking for a trend, we want to end up with pairs of x and y values. One way we might proceed is to put a number of individually marked males into a sand-filled arena and record all encounters over, say, 20 min, noting the males involved and the highest level (on our scale of 1 to n) to which each encounter progressed. One problem with this approach, however, is that some males would interact more than once. The pairs of x and y values arising from each repeat encounter could not be used independently because body size ratio would be confounded with pair of opponents and escalation levels might be influenced by the males' past experience of each other. It would therefore be better to arrange encounters between different pairs of males to provide independent replicates of a range of size ratios. Since we are using ordinal (rank) measures of y and selected ratios as x, we should test for significance in our trend using a Spearman rank correlation rather than regression.

REFINING HYPOTHESES AND PREDICTIONS

Depending on the results of our experiments or observations, a prediction may be borne out and support the hypothesis or it may not. Failure to bear out a prediction, of course, does not necessarily sound the death knell for the hypothesis. Predictions often reflect specific interpretations of a hypothesis. For instance, take Prediction 3B from the stickleback example. This predicts a greater rate of food intake by fish in larger groups on the hypothesis (3B) that fish can afford to spend less time vigilant (and thus more time looking for food) in bigger groups. Suppose a suitable experiment failed to support the prediction: there was no significant increase in intake rate with group size. Does this undermine the hypothesis? The answer, clearly, is no. The hypothesis is concerned with a positive relationship between group size and time available to look for food. Finding **more** food in a given time is one prediction that can be derived from it, but another is that, with more time on their hands, fish can afford to be more selective about what they eat. Under this prediction, intake rate may not change with group size but the **range** of items taken should. Thus fish might tend to select larger or healthier-looking *Daphnia* as group size increases. If such variation in the quality of *Daphnia* has been scrupulously controlled for in the experiment testing Prediction 3B, of course, the hypothesis would be in greater difficulty.

By contrast, let's reconsider Prediction 2B (see Chapter 2). We can use this to make two points. The first is that predictions deriving logically from a hypothesis do not always provide a rigorous test of the hypothesis. Prediction 2B predicts that chicks will find food faster if they forage with companions, an expectation derived from Hypothesis 2B which assumes chicks use the behaviour of companions to help locate food. Suppose the experiment we designed bore out the prediction with a significant difference between 'solitary' and 'companion' groups in the number of rice grains eaten per unit time. Would this increase our faith in the hypothesis? 'Yes, but not a lot' is probably the answer. The hypothesis is specific about the form of feeding advantage conferred by companions – they provide information; the prediction merely tests a general consequence of this – chicks will find food faster. Unfortunately, companions could cause chicks to find food faster in ways other than providing information, e.g. by making them feel safer and thus spend less time vigilant, like the sticklebacks. A more rigorous prediction from Hypothesis 2B would thus stipulate an increase in feeding rate as a **result of moving closer to successful companions.**

The second point we can make from Prediction 2B is therefore that the results of testing a prediction, whether supportive or contradictory, can lead to a refinement of the hypothesis. Suppose, once again, that chicks did find food faster with companions. To dispel our niggling doubt about the 'safety in numbers' effect, we could conduct a brief additional experiment. If we placed subjects and companions on different sides of a clear partition, so they could see each other but not intermix, any increase in the subjects' rate of food intake could not be due to searching near successful companions. If we sunk food in wells below the surface of the arena, we could also rule out remote observational learning about the kind of object at which to peck. Of course, we have now drifted into testing an alternative hypothesis (that companions provide security through a safety in numbers effect). If subjects still showed increased intake with companions, the 'safety in numbers' hypothesis would start to look a better bet than the original 'feeding information' hypothesis so we may wish to switch to this and develop further testable predictions.

Although we have presented hypotheses and their predictions in a rather cut-and-dried fashion through this book, it is clear that there is really considerable fluidity in both. The relationship between hypothesis, prediction and test is a dynamic one and it is through the modifying effects of each on the others that science proceeds.

Summary

1. Predictions derived from hypotheses dictate the experiments and observations needed to test hypotheses. As a result of testing, hypotheses may be rejected, provisionally accepted or modified to generate further testable hypotheses.

2. Decisions about experimental/observational measurement and the confirmatory analysis of such measurements are inter-dependent. The intended analysis determines very largely what should be measured and how, and should thus be clear from the outset of an investigation.

3. Some kind of yardstick is needed in confirmatory analysis to allow us to decide whether there is a convincing difference or trend in our measurements (i.e. whether we can reject the null hypothesis of no difference or trend). The arbitrary, but generally accepted, yardstick is that of statistical significance. Significance tests allow us to determine the probability that a difference or trend as extreme as the one we have obtained could have occurred purely by chance. If this probability is less than an arbitrarily chosen threshold, usually 5%, but sometimes 1% or 10%, the difference or trend is regarded as significant and the null hypothesis is rejected.

4. Different significance tests may demand different attributes of the data. Parametric and nonparametric tests differ in the

assumptions they make about the distribution of data values within samples and the kinds of measurement they can cope with. Tests can also be used in specific/one-tailed or general/two-tailed forms depending on prior expectations about the direction of differences or trends.

5. Basic tests for a difference include chi-squared, the Mann–Whitney U test and one- and two-way analysis of variance. Each has a number of requirements which must be taken into account and care is needed not to make multiple use of two-group difference tests in comparing more than two groups of data.

6. Basic tests for a trend include Spearman rank correlation and regression analysis. Correlation is used when we merely wish to test for an association between two variables. Regression is used when we experimentally change the values of one variable and observe the effect on another to test for a cause and effect relationship between two variables. Regression analysis yields more quantitative information about trends than correlation but makes more stringent demands on the data.

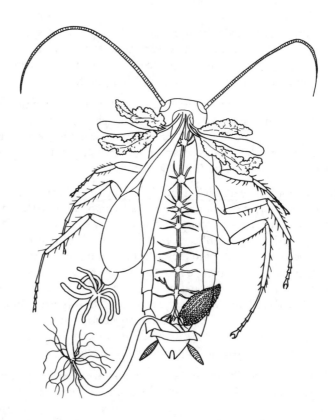

7. Testing predictions requires careful thought about methods of measurement, experimental/observational procedure, replication and controlling for confounding factors and floor and ceiling effects.

REFERENCES

Bailey, N. T. J. (1981) *Statistical methods in biology*, 2nd edition. Hodder and Stoughton, London.

Martin, P. and Bateson, P. (1986) *Measuring behaviour*. Cambridge University Press, Cambridge.

Meddis, R. (1984) *Statistics using ranks: a unified approach*. Blackwell, Oxford.

Sokal, R. R. and Rohlf, F. J. (1981) *Biometry*, 2nd edition. Freeman, San Francisco.

4

PRESENTING INFORMATION

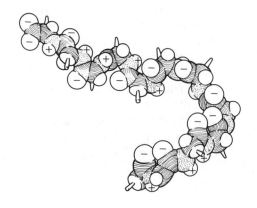

If your investigation has gone to plan (and possibly even if it hasn't), you will have generated a pile of data which somehow needs to be presented as a critical test of your hypothesis. Performing appropriate significance tests is only one step on the way. While significance tests will help you decide whether a difference or trend is interesting, this information still has to be put across so that other people can evaluate it for themselves. There are two reasons why we should take care how we present our results. The first is to ensure we get our message over; there is little point making a startling discovery if we can't communicate it to anyone. The purpose of our investigation was to test a hypothesis. We might conclude that the results support the hypothesis or that they undermine it. Whichever conclusion we reach we must sell it if we wish it to be taken seriously. Since scientists are by training sceptical, selling our conclusion may demand some skilful presentation and marshalling of arguments. The second reason is that we must give other people a fair chance to judge our conclusions. As we saw earlier, simply saying that some difference of trend is significant doesn't tell us how strong the effect is. It is important to present results in such a way that others can make up their own minds about how well they support our conclusions. In this section, we shall look at some conventions in presenting information that help satisfy both these requirements. We begin with some simple points about figures and tables.

PRESENTING FIGURES AND TABLES

We stressed earlier that it is usually not helpful to present raw data. Raw data are often too numerous and the information in them too difficult to assimilate for useful presentation. Instead, we summarize them in some way and present the summary form. We have already dealt with summary statistics in a general way when we discussed exploratory analysis; here we discuss their use in presenting the results of confirmatory analysis.

Although it is usually obvious that raw data require summarizing, there can be a temptation to summarize everything, as if summary statistics or plots were of value in themselves. In confirmatory analyses, they are of value only to the extent that they help us evaluate tests of hypotheses. We thus need to be selective in distilling our results. Naturally, the summaries and forms of presentation that are most appropriate will depend on the type of confirmatory analysis. It is therefore easiest to deal with different cases in turn.

Presenting analyses of difference

Where we are dealing with analyses of difference, the important information to get across is a summary of the group values being compared. The form this takes will vary with the number of groups and levels of grouping involved. There are two basic ways of presenting a summary of differences: figures and tables. As we have argued previously, figures tend to be easier to assimilate than tables, even when the latter comprise summary statistics. However, tables may be more economical when large numbers of comparisons are required, or where comparisons are subsidiary to the main point being argued but helpful to have at hand. If tables **are** used, it is important that they present **all** the key summary information necessary to judge the claims they make. This usually means (a) summary statistics (e.g. means ± standard errors*, medians ± confidence limits) for each of the groups being compared, (b) the sample size (n) for each group, (c) test statistic values, (d) the probabilities (p-values) associated with the test statistic values and (e) an explanatory legend detailing what the table tells us. The

---

* While there are various forms of summary statistic, means ± standard errors are widely used because the mean of a set of values is an easy concept to grasp and because the standard error estimates the distribution of **means** within the population which approaches a normal distribution as the sample size increases. It is thus usually legitimate to quote means ± standard errors as summary statistics even when the distribution of data **values** demands a nonparametric significance test.

test statistics and p-values can be presented either in the table itself or in the legend. The same information, of course, should be presented in figures except that the summary statistics are represented graphically (e.g. as bar charts) instead of as numbers, and information about sample sizes, test statistics and probability levels more conventionally goes in the legend (now usually called the figure caption) rather than in the figure itself. (Nevertheless, as long as it doesn't clutter the figure and detract from its impact, it can be very helpful to include statistical information within the figure and we shall do this later where appropriate.)

Differences between two or more groups (with one level of grouping)

Here we are presenting the kinds of result that might emerge from a Mann–Whitney U test, or a one-way analysis of variance. Suppose we have tested for a difference in growth rate (general prediction) between two groups of plants, one given a gibberellin (growth-promoting) hormone treatment, the other acting as an untreated control. The treated group contained 12 plants and the untreated group eight. Using a one-way analysis of variance for two groups, we discover a significant difference at the 0.1% ($p < 0.001$) level between the groups, with treated plants growing to a mean ($\pm$ standard error) height of 14.75 ± 0.88 cm during the experimental period, and controls growing to a mean height of 9.01 ± 0.63 cm. We could present these results as in Table 4.1a. Note the legend explaining exactly what is in the table. The

Table 4.1a The mean height to which plants grew during the experimental period when treated with gibberellin or left untreated

| | Experimental groups | | |
	Treated	Untreated	Significance
Mean ($\pm$ s.e.) height (cm)	14.75 ± 0.88	9.01 ± 0.63	$H = 12.22, p < 0.001$
n	12	8	

significance column could be omitted from the table, in which case the test statistic and probability level should be given in the legend. The legend would now read:

Table 4.1a The mean height to which plants grew during the experimental period when treated with gibberellin or left untreated. H comparing the two groups $= 12.22$, $p < 0.001$

An alternative and frequently adopted convention in presenting significance levels is to use asterisks instead of the test statistic

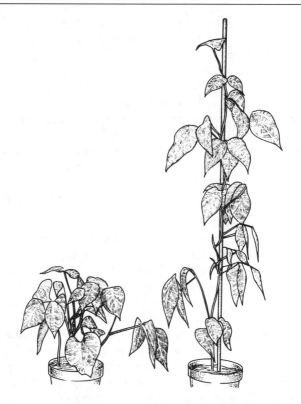

and probability numbers. In this case, different levels of probability are indicated by different numbers of asterisks. Usually * denotes $p < 0.05$, ** $p < 0.01$ and *** $p < 0.001$, but this can vary between investigations so it is important to declare your convention when you first use it. Using the asterisks convention, the table would now read as in Table 4.1b. Exactly the same forms of presentation, of course, could be used for comparisons of more than two groups. (However, if we had used a U test to test for a difference between two groups, we should now have to change to a one-way analysis of variance to avoid abuse of a two-group difference test (see Chapter 3).)

Table 4.1b The mean height to which plants grew during the experimental period when treated with gibberellin or left untreated. ***, $H = 12.22$, $p < 0.001$

	Experimental groups		
	Treated	Untreated	Significance
Mean ($\pm$ s.e.) height (cm)	14.75 $\pm$ 0.88	9.01 $\pm$ 0.63	***
n	12	8	

Table 4.2a The number of plants surviving treatment with different herbicides

	Experimental group				
	Herbi-cide 1	Herbi-cide 2	Herbi-cide 3	Control	Significance
Number of plants surviving	15	8	6	27	$\chi^2 = 19.3, p < 0.001$

The presentation of means and standard errors (or medians and confidence limits) is appropriate whenever we are dealing with analyses that take account of the variability within data samples. In chi-squared analyses, however, where we are comparing simple counts, there is obviously no variability to represent. If we were presenting a chi-squared analysis of the number of plants surviving each of three different herbicide treatments and one control treatment, therefore, the table would be as shown in Table 4.2a. Table 4.2b shows an alternative presentation. If we wish to present our results as figures rather than tables, we can convey the same information using simple bar charts. Thus Table 4.1a can be recast as Fig. 4.1a. Similarly, Table 4.1b could be recast as Fig. 4.1b. For a comparison of three groups, say comparing the effectiveness of the lambda bacteriophage in killing three strains of *Escherichia coli* suspected of differing in susceptibility, the figure might be as shown in Fig. 4.2a.

In Figs 4.1a, b and 4.2a, we have assumed a one-way 'analysis of variance' was used to test a **general** prediction (hence the test statistic H). If we had instead tested a **specific** prediction because we had an *a priori* reason for expecting a rank order of effect (e.g. gibberellin-treated plants would grow taller than untreated plants

Table 4.2b The number of plants surviving treatment with different herbicides. ***, $\chi^2 = 19.3, p < 0.001$

	Experimental group				
	Herbi-cide 1	Herbi-cide 2	Herbi-cide 3	Control	Significance
Number of plants surviving	15	8	6	27	***

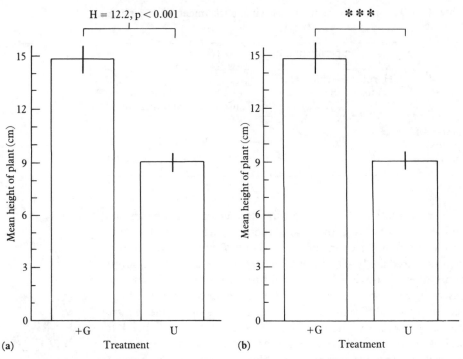

Fig. 4.1a and b The mean height to which plants grew during the experimental period when treated with gibberellin (+G, $n = 12$) or left untreated (U, $n = 8$). ***, $H = 12.2$, $p < 0.001$. Bars represent standard errors.

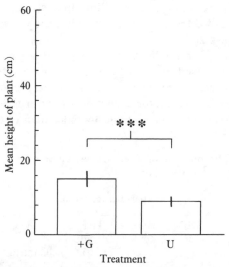

Fig. 4.1b (different scales) The mean height to which plants grew during the experimental period when treated with gibberellin (+G, $n = 12$) or left untreated (U, $n = 8$). ***, $H = 12.2$, $p < 0.001$. Bars represent standard errors.

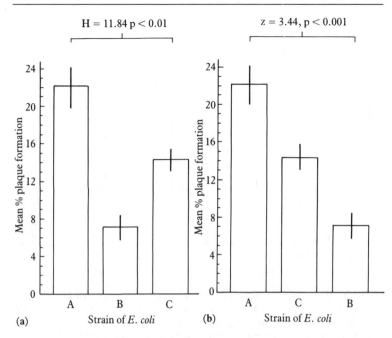

Fig. 4.2 The mean percentage area of plaque (= bacterial death) formation by lambda bacteriophage on three strains (A–C) of *E. coli*. Bars represent standard errors. a Nonparametric analysis of variance testing a general prediction of difference between strains. $N = 8$ cultures in each case. b Nonparametric analysis of variance testing a specific prediction of difference between strains (A>C>B). $N=8$ cultures in each case.

(Fig. 4.1), or strain B of *E. coli* would be most resistant and strain A least resistant to attack by the phage (Fig. 4.2)), we should recast the figures with groups in the predicted order and quote the test statistic z rather than H. Thus Fig. 4.2a could be recast as Fig. 4.2b.

For most people, bar charts like these convey the important differences between groups more clearly and immediately than equivalent tables of numbers. However, it is worth stressing some key points which help to maximize the effectiveness of a figure.

1. Make sure the scaling of numerical axes is appropriate for the difference you are trying to show. For instance, the impact of Fig. 4.1b is much reduced by choosing too large a scale (see foot of page 88).

2. Always use the **same** scaling on figures that are to be compared with one another. Thus, Fig. 4.3a, b is misleading because the different scaling makes the magnitude of the bars look the same in (a) and (b). Using the same scale, as in Fig. 4.3c, shows that there is in fact a big difference between (a) and (b).

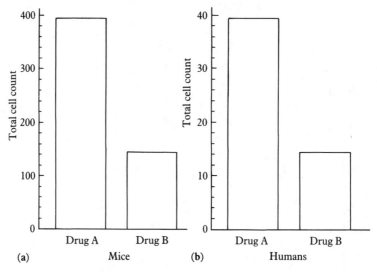

Fig. **4.3a** The total number of T-helper cells in experimental samples from laboratory mice following administration of two different cytotoxic drugs (A and B). $N = 40$ samples for each drug treatment.
Fig. **4.3b** As **4.3a** but for experimental samples from humans.

3. Make sure axes are numbered and labelled properly and that labels are easy to understand and indicate the units used. Avoid obscure abbreviations in axis labels: these can easily be ambiguous and misleading or unnecessarily difficult to interpret.

4. Axes do not have to start at zero. Presentation may be more economical if an axis is broken and starts at some other value. Thus, Fig. 4.4a could be recast as Fig. 4.4b with the break in the vertical axis indicated by a double slash.

5. Include indications of variability (standard errors, etc.) where appropriate. Also include sample sizes and p-values as long as these don't clutter the figure. If they do, put them in the legend.

6. Always provide a full, explanatory legend. The phrasing of the legend should be based on the prediction being tested and the legend should include any statistical information (sample sizes, test statistics, etc.) not included in the figure (do not repeat information in both figure and legend, though). The legend should allow a reader to assess the information in the figure without having to plough through accompanying text to find more detailed discussion.

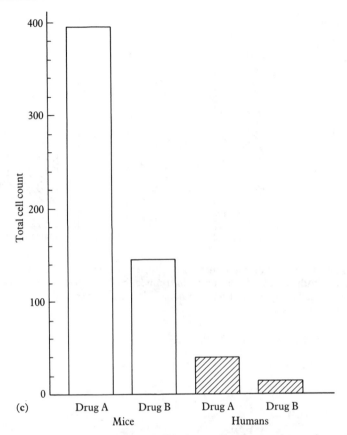

Fig. 4.3c The total number of T-helper cells in experimental samples from laboratory mice (open bars) and humans (shaded bars) following administration of two different cytotoxic drugs (A and B). $N = 40$ samples for each drug treatment and species.

Differences between two or more groups (with more than one level of grouping)

Here we are concerned with the sort of results that might arise from a two-way analysis of variance or $n \times n$ chi-squared analysis. Presentation is now a little trickier because of the number of comparisons we need to take into account. A table is probably the simplest solution. For instance, suppose we had carried out a two-way analysis of variance looking at the difference in the frequency of accidental egg damage between three strains of battery hen maintained in three different housing conditions. Here we have two levels of grouping (strain and housing condition) with three groups at each level. The analysis tests for a difference between strains (controlling for housing condition), a difference

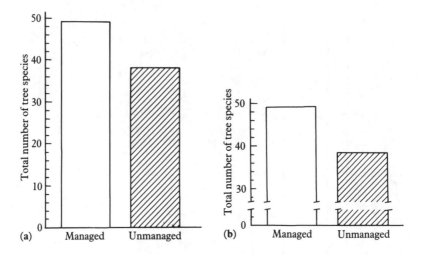

Fig. 4.4a and b The total number of species of tree bearing epiphytes in 2 km² study areas of rainforest in Bolivia where forests are managed economically (open bar) and unmanaged (shaded bar). $N = 1 \times 2$ km² area in each case.

between housing conditions (controlling for strain) and any interaction between the two levels of grouping (see Chapter 3). The best way to present the differences between groups within levels is to tabulate the summary statistics for each of the nine (3 × 3 groups) cells and include the test statistics in the legend. Thus testing for **any** difference between groups (i.e. not predicting a difference in any particular direction) might give the results shown in Table 4.3.

This analysis reveals significant effects of both strain and housing conditions on egg breakage. These are obvious from the summary statistics in the table: breakage in strains 1 and 3 is relatively high under housing conditions A and B but drops sharply in condition C. In contrast, breakage in strain 2 is highest in conditions B and C and lowest in A. Damage tends to be greater in types A and B housing than in type C. In addition to these **main effects**, however, there is also a significant interaction between strain and housing condition (see legend to Table 4.3) with the effect of housing differing between strains. Although interaction effects can also be gleaned from a table of summary statistics like Table 4.3, they can be presented more effectively as a figure; one of the levels of grouping constitutes the x-axis and the measure being analysed is the y-axis. The relationship between the measure and the x-axis grouping can then be plotted for each group in the second level. Figure 4.5 shows such a plot for the interaction in Table 4.3. The lines in the figure, of course, simply indicate the groups of data: they are in no way comparable with statistically fitted lines. Full details of the analysis are given in the legend because such a figure would not normally be presented as well as the summary table since it repeats information already given in the table. From Fig. 4.5 it is clear that, while all three

Table 4.3 The mean (± s.e.) percentage number of eggs broken during the experimental period by three strains of battery hen (1–3) under three different housing conditions (A–C). Nonparametric two-way analysis of variance shows a significant effect of both strain ($H = 8.84$, $p < 0.05$) and housing ($H = 10.12$, $p < 0.01$) and a significant interaction between the two ($H = 14.92$, $p < 0.01$). $n = 4$ in each combination of strain and housing condition

| | | Strain | | |
		1	2	3
	A	47.50 ± 2.95	0.50 ± 0.50	23.75 ± 1.49
Housing condition	B	43.00 ± 1.47	13.50 ± 1.19	18.25 ± 0.48
	C	5.50 ± 1.44	11.75 ± 0.85	4.50 ± 0.65

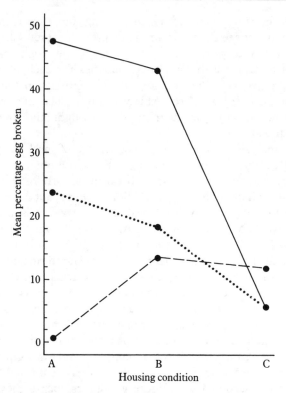

Fig. 4.5 The mean percentage number of eggs broken by three strains of battery hen (solid, strain 1; dashed, strain 2; dotted, strain 3) in three different housing conditions (A–C). Nonparametric two-way analysis of variance showed a significant effect of both strain ($H = 8.84$, d.f. $= 2$, $p < 0.05$) and housing condition ($H = 10.12$, d.f. $= 2$, $p < 0.01$) and a significant interaction between the two ($H = 14.92$, d.f. $= 4$, $p < 0.01$). $N = 4$ in each combination of strain and housing condition.

strains show differences in egg damage across housing conditions, the direction and degree of decline are different in different strains. This implies that the effect of housing condition varies with strain (which is what is meant by an interaction between housing condition and strain).

With an $n \times n$ chi-squared analysis, we are just dealing with total counts in each cell so there are no summary statistics to calculate and present. The simplest presentation is thus an $n \times n$ table with each cell containing the observed and expected values for the particular combination of groups (the expected value in each cell usually goes in brackets). Table 4.4 shows such a presentation for a chi-squared analysis of the effects of temperature and soil type on the number of seeds out of 150 germinating in a seed tray.

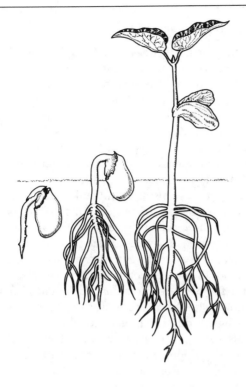

Table 4.4 The number of seeds germinating in a tray in relation to temperature (low, 5 °C; high, 25 °C) and soil type. Expected values in brackets. $\chi^2 = 14.38$, d.f. = 1, $p < 0.001$

	Number of seeds germinating	
	In clay soil	In sandy soil
Low temperature	40 (57.97)	100 (82.03)
High temperature	131 (113.03)	142 (159.97)

Presenting analyses of trends

Presenting trend analyses is rather simpler because, in most cases, a scattergram with or without a fitted line is the obvious format. When it comes to more complicated trend analyses that deal with lots of different measures at the same time, summary tables rather than figures may be necessary. However, these need not concern us here.

Presenting a correlation analysis

Since correlation analysis does not fit a line to data points, presentation consists simply of a scattergram, though depending on how we have replicated observations this may include some summary statistics (see below). Information about test statistics, sample sizes and significance could be given in the figure, but it is more usual to include it in the legend. Thus Fig. 4.6a shows a plot of the number of food items obtained by male house sparrows in relation to their dominance ranking with other males in captive flocks of six (rank 1 is the most dominant male that tends to beat all the others in aggressive disputes and rank 6 is the least dominant that usually loses against everyone else). In this case, observations were repeated for three sets of males so there are three separate points (y-values) for each x-value in the figure.

Although there is a significant trend towards dominant males getting more food, the correlation is negative because we chose to use a rank of 1 for the most dominant male and a rank of 6 for the least dominant. Rankings are frequently ordered in this way, leading to the slightly odd situation of concluding a positive trend (e.g. dominants get more food) from what looks like a negative trend (the number of food items decreases with increasing rank number). There is no reason, of course, why dominance shouldn't be ranked the other way round (6 = most dominant, 1 = least dominant) so that a positive slope actually appears in the figure.

Sometimes when replicated observations are presented in a scattergram, they are presented as a single mean or median with appropriate standard error or confidence limit bars. Thus an alternative presentation of Fig. 4.6a is shown in Fig. 4.6b. Note that a different explanatory legend is now required because the figure contains different information.

In some cases, replication may not occur throughout the data set. Say we decided to sample a population of minnows in a stream to see whether big fish tended to have more parasites. To avoid the difficulties of making accurate measurements of fish size in the field and possibly injuring the fish, we visually assess those we catch as belonging to one of six size classes. We then count the signs of parasitism on them and return them to the water. Because we have no control over the number of each size class we catch, we end up with more samples for some classes than for others. When we come to present the data, we could present them as individual data points for each size ranking (Fig. 4.7a) or condense replicated data for ranks to means or medians (Fig. 4.7b). In the latter case, only some points might have error or confidence limit bars attached to them because only some ranks are replicated (see Fig. 4.7a). Data for those ranks that are not replicated are still presented as single points.

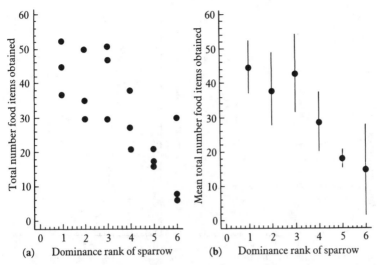

Fig. 4.6a The number of food items obtained during the period of observation by male house sparrows of different dominance status in groups of six (rank 1, most dominant; rank 6, least dominant, data for three groups at each rank). $r_s = -0.77$, $n = 18$, $p = 0.001$.

Fig. 4.6b The mean number of food items obtained during the period of observation by male house sparrows of different dominance status in groups of six (rank 1, most dominant; rank 6, least dominant). $r_s = -0.77$, $n = 18$, $p = 0.001$. Bars represent standard errors.

It is, of course, important to remember that even where correlations are presented as mean or median values rather than independent data points, the correlation analysis itself (i.e. the calculation of the correlation coefficient) is still performed on the independent data points, not on the means or medians. Values of n are thus the same in Figs 4.7a and 4.7b. Correlations can be performed on summary statistic values, but obviously a lot of information is lost from the data and n-sizes are correspondingly smaller.

Presenting a regression analysis

Presenting a regression analysis is essentially similar to presenting a correlation except that a line needs to be fitted through the data points. If the trend isn't significant, so that a line should not be fitted, a figure probably isn't necessary in the first place. The details of calculating a regression line have been given earlier. You may sometimes come across regression plots which show confidence limits as curved lines above and below the regression line itself. However, we shall not be dealing with these here.

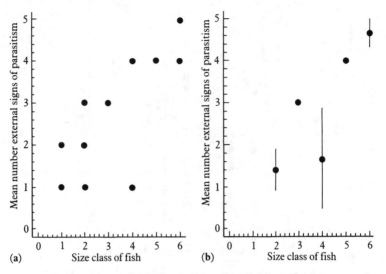

Fig. 4.7a The relationship between the size of minnows (arbitrary size classes) and the number of signs of parasitic infection observed on them. $r_s = 0.74$, $n = 11$, $p = 0.019$. b The relationship between the mean size of minnows (arbitrary size classes) and the number of signs of parasitic infection observed on them. $r_s = 0.74$, $n = 11$, $p = 0.019$. Bars represent standard errors.

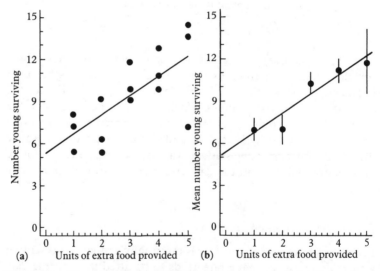

Fig. 4.8a The number of chicks surviving to their first winter in relation to the number of units of extra food provided during the breeding season in three populations of moorhen. $F = 13.27$, d.f. $= 1, 13$, $p < 0.01$. b The mean number of chicks surviving to their first winter in relation to the number of units of extra food provided during the breeding season in three populations of moorhen. $F = 13.27$, d.f. $= 1, 13$, $p < 0.01$. Bars represent standard errors.

As with correlation, data can be presented as independent points or, where replicated for particular x-values, as means or medians. Once again, where means or medians are presented, significance testing and the fitting of the line are still done using the individual data points, not the summary statistics. Figure 4.8 presents a regression of the effect of additional food during the breeding season on the number of young moorhens surviving into their first winter in three study populations. Five different quantities of food were used and the three populations received them in a different order over a 5-year experimental period. In Fig. 4.8a, the number for each population are presented separately; in Fig. 4.8b they are presented as means ($\pm$ s.e.) across the three study populations.

PRESENTING RESULTS IN THE TEXT

So far in this section, we have assumed that results will be presented as figures or tables. Figures and tables, however, take up a lot of space in a report and may not be justified if the result is relatively minor or there is a strict limit on the length of the report. In such cases, analyses can be summarized in parentheses in the text of the Results section (see later). The usual form for a difference analysis is to quote the summary statistics, test statistic, sample size or degrees of freedom and p-value. Thus the information in Table 4.1a could easily be presented in the text as:

Treatment with gibberellin resulted in a significant increase in growth compared with non-treated controls (mean ($\pm$ s.e.) height of treated plants = 14.75 ± 0.88 cm, $n = 12$; mean height of controls = 9.01 ± 0.63 cm, $n = 8$; $H = 12.22$, $p < 0.001$).

For a trend, it is usual to quote just the test statistic, sample size or degrees of freedom and the p-value. For instance, the information in Fig. 4.8a could be summarized as follows:

The number of chicks hatching during a breeding season that survived into their first winter increased significantly with the amount of extra food provided within the population ($F = 13.27$, d.f. = 1, 13, $p < 0.01$).

It is impossible to generalize about when an analysis could be presented in the text rather than in separate figures or tables. Sometimes, as we have said, it is simply a matter of limited space. However, rough guidelines might include the following: (a) difference analyses between only two or three groups; (b) corroborative analysis, supporting a main analysis already presented as a figure or table (for instance, if a main analysis showed a significant correlation between body size and fighting ability, a corroborative analysis might check that body size was not confounded with age and that the correlation could not have arisen because bigger individuals had more experience of fighting); (c) analyses providing background information (e.g. showing a significant sex difference in body size where this is germane to, say, an analysis of the diet preferences of the two sexes).

WRITING REPORTS

Just as figures and tables of data should be presented properly to ensure they are effective, so care must be taken in the text of a report. In the scientific community, reports of experiments and observations are usually published in the form of papers in professional journals, or sometimes as chapters in specialist books. In all cases, however, the aim is both to communicate the findings of a piece of research **and** provide the information necessary for someone else to repeat the work and check out the results for themselves. Both these elements are crucial and, as a result, scientific papers are usually refereed by other people in the same field to make sure they come up to scratch before being published. Not surprisingly, a more or less standard format for reports has emerged which divides the textual information into well-recognized sections that researchers expect to see and know how to refer to to find out about different aspects of the work. Learning to use this format properly is one of the most important goals of any basic scientific training. To finish off, therefore, we shall discuss the general structure of a report and what should and should not go in each of its sections; then we shall develop a full report, incorporating our various points about text and data presentation, from some of our main example observations.

The sections of a report

There are five principal sections in a report of experimental or observational work: **Introduction, Methods, Results, Discussion** and **References**. Sometimes it is helpful to have some small additional sections such as **Abstract, Conclusions** and **Appendices**, but we shall deal with these later.

Introduction

The Introduction should set the scene for all that follows. Its principal objective is to set out: (a) the **background** to the study, which means any theoretical or previous experimental/observational work that led to the hypotheses under test. Background information is thus likely to include references to previously published work and sometimes a critical review of competing ideas or interpretations. It might also include discussion about the timing (seasonal, diel, etc.) of experiments/observations and, in the case of fieldwork, the reasons for choosing a particular study site; (b) a clear statement of the hypotheses and predictions that are being tested; and (c) the rationale of the study, i.e. how its design allows the specified predictions to be tested and alternatives to be excluded. The Introduction should thus give the reader a clear idea as to why the study was carried out and what it aimed to investigate. The following is a brief example:

> Reptiles are ectotherms and thus obtain most of the heat used to maintain body temperature from the external environment (e.g. Davies, 1979). Rattlesnakes (*Crotalus* spp.) do this by basking in the sun or seeking warm surfaces on which to lie (Bush, 1971). An increased incidence of snake bites in the State over the past two years has been attributed to a number of construction projects that have incidentally provided rattlesnakes with concrete or tarmac surfaces on which to bask (North, 1989). The aim of this investigation was to study the effect of the construction projects on basking patterns among rattlesnakes to see whether these might increase the exposure of people to snakes and thus their risk of being bitten. The study tests two hypotheses: (a) concrete and tarmac surfaces are preferred basking substrates for rattlesnakes and (b) such surfaces result in a higher than average density of snakes near humans.

When the reader moves on to the Methods and Results sections, he/she will then appreciate why things were done the way they were. Reading Methods or Results sections without adequate introductory information can be a frustrating and often fruitless business since the design of an experiment or observation usually makes sense only in the context of its rationale. As we shall see below, it is sometimes appropriate to include background material in the Discussion section, but in this case it should be to help develop an interpretation or conclusion, not an afterthought about information relevant to the investigation as a whole; if it is the latter, it should be in the Introduction.

Methods (or Materials and Methods)

The Methods section is perhaps the most straightforward. Nevertheless, there are some important points to bear in mind.

Chief among them is providing enough detail for someone else to be able to repeat what you did **exactly**. Clearly, the precise detail in each case will depend on the investigation, but points that need attention are likely to include the following.

• *Experimental/observational organisms or preparations.*   The species, strain, number of individuals used, housing conditions and husbandry, age and sex, etc. for organisms; the derivation and preparation and maintenance techniques etc. for preparations (e.g. cell cultures, histological preparations, pathogen inoculations).

• *Specialized equipment.*   Details (make, model, relevant technical specifications, etc.) of any special equipment used. This usually means things like tape or video recorders, spectrometers, oscilloscopes, automatic data-loggers, optical equipment such as telescopes, binoculars or specialized microscopes, centrifuges, respirometers, specially constructed equipment such as partitioned aquaria, choice chambers, etc. Run-of-the-mill laboratory equipment like glassware, balances, hotplates and so on don't usually require details, though the dimensions of things like aquaria or other containers used for observation and the running temperature of heating devices, etc. should be given.

• *Study site (field work).*   Where an investigation has taken place in the field, full details of the study site should normally be given. These should include its location (e.g. grid reference) and a description of its relevant features (e.g. size, habitat structure, use by man) and how these were used in the investigation.

• *Data collection.*   This should include details of all the important decisions that were made about collecting data. Again, it is impossible to generalize, but the following are likely to be important in many investigations: any pretreatment of material before experiments/observations (e.g. isolation of animals, drug treatment, surgical operations, preparation of cell cultures, staining); details of experimental/observational treatments **and controls**; sample sizes and replication; methods of measurement and timing; methods of recording (e.g. check sheets, tape recording, tally counters, etc.); duration and sequencing of experimental/observational periods; details of any computer software used in data collection. Of course, it is important not to go overboard. For instance, it isn't necessary to relate that a check sheet was ticked with a red ball-point pen rather than a black one, but if the pen was used to stimulate aggression in male sticklebacks (which often attack red objects) then it would be relevant to state that a ball-point pen was used and that it was red.

Results

The Results section is in some ways the most difficult to get right. Many students regard it as little more than a dumping ground for all manner of summary and, worse, raw data. Explanation, where it exists at all in such cases, frequently consists of an introductory 'The results are shown in the following figures ...' and a terminal 'Thus it can be seen ...'. A glance at any paper in a journal will show that a Results section is much more than this. At the other extreme, explanation within the Results often drifts into speculative interpretation which is more properly the province of the Discussion (see below).

A Results section should do two things and **only** two things: first, it should present the data (almost always in some summarized form, of course) necessary to answer the questions posed; and second, it should explain and justify the analytical approach taken so that the reasons for choice of test and modes of data presentation are clear. The section should thus include a substantial amount of explanatory text, but explanation should be geared solely to the analyses and presentation of data and not the interpretations or conclusions that might be inferred from them. An example might be as follows:

> Figure 1 shows that rattlesnakes are significantly more likely to be found on concrete (Figure 1a) and tarmac (Figure 1b) surfaces around dawn and dusk than around midday. Since many of the construction projects in the survey of snake bite incidence have involved highways (Greenbaum *et al.*, 1984), this temporal pattern of basking may result in highest snake/human encounter at times when public conveniences are closed and motorists are forced to relieve themselves at the roadside. Indeed Table 1 shows a strong association for three highways between time of day and number of motorists stopping by the roadside.

It is also important that **all** the analyses and presentations of data involved in the report appear in the Results section (as figures, tables or in the text) and only in the Results section; no analysis should appear in any other section.

Discussion

The Discussion is the place to comment on whether the results support or refute the hypotheses under test and how they relate to the findings of other studies. The Discussion thus involves interpretation and reasonable speculation, with further details about the material investigated and any corroborative/contradictory/background information as appropriate. As we have said, however, while the Discussion may flesh out, comment, compare and conclude, it should not bring in new analysis. Neither should it develop background information that is more appropriate to the Introduction (see earlier). The kind of thing we'd expect might be as follows:

> The results suggest that concrete and tarmac surfaces are not favoured for basking by rattlesnakes in comparison with broadly equivalent natural surfaces when relative area is taken into account. One reason for this might be the greater proximity and greater density of cover close to the natural surfaces sampled. Many snakes (Jones, 1981), including rattlesnakes (Wilson, 1976), prefer basking areas within easy escape distance of thick cover. Despite not being preferred by snakes, the greater incidence of bites on concrete and tarmac surfaces can be explained in terms of the greater intensity of use of these surfaces by humans. However, Wilson (1976) has noted that the probability of attack when a snake is encountered increases significantly if there is little surrounding cover. The paucity of cover around the concrete and tarmac samples may thus add to the risk of attack in these environments.

References

Your report should be referenced fully throughout, with references listed chronologically in the text and alphabetically in a headed Reference section at the end. References styles vary enormously between different kinds of report so there is no one accepted format. However, a style used very widely is illustrated below and we suggest using it except where you are explicitly asked to adopt a different style. In this style, references in the text should take the form:

> ... Smith (1979, 1980) and Grant *et al.* (1989) claim that, during a storm, a tree 10 m in height can break wind for over 100 m (but see Jones and Green, 1984; Nidley, 1984, 1986). ...

In the References list at the end, journal references take the form:

Grant, A. J., Wormhole, P. and Pigwhistle, E. G. (1989) Tree lines and the control of soil erosion. Int. J. Arbor. 121, 42–78.
Jones, A. B. and Green, C. D. (1984) Soil erosion: a critical review of the effect of tree lines. J. Plant Ecol. 83, 101–107.
Smith, E. F. (1979) Planting density and canopy size among deciduous trees. Arbor. Ecol. 19, 27–50.

Smith, E. F. (1980) Planting density and growth rate among deciduous trees. Arbor. Ecol. 20, 38–52.

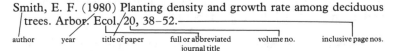

| author | year | title of paper | full or abbreviated journal title | volume no. | inclusive page nos. |

for books they take the form:

Nidley, R. (1984) Deforestation and its impact on national economies.   Hacker Press, London.

title of book underlined   publisher   place of publication

and for chapters in edited volumes the form:

Nidley, R. (1986) Economic growth and deforestation. In Sustainable economics and world resources, eds A. B. Jones and C. D. Green, pp. 64–78. Hacker Press, London.

Where more than one source by a particular author (or set of authors) in a particular year is referred to, the sources can be distinguished by using lower case letter suffixes, e.g. (Smith, 1976a, b) indicates that you are referring to two reports by Smith in the year 1976. The order in which you attribute a, b, c, etc. is determined by the order in which you happen to refer to the publications in your report, not the order in which they were published in the relevant year.

• *Personal observations and personal communications.* Although most of the references you will make will be to work by other people, or yourself, that has been published in some form, it is occasionally appropriate to refer to unpublished observations. This usually arises where some previous, but unpublished, observation is germane to an assumption, fact, technique, etc. that you are relying on in your own report. If such observations are your own, they can be referred to in the text as '(personal observation)' or '(pers. obs.)'. If they have been reported to you by someone else, then they can be referred to as, for example, '(P. Smith, personal communication)' or '(P. Smith, pers. comm.)' – note that the name of the person providing the information is given as well.

Other sections of a report

In some cases, there may be additional sections to a report.

• *Abstract*. This is a short summary of the aims and major findings of your investigation. The idea is to provide the reader with a quick overview of what you've done and what is interesting about it so that he/she can decide whether they want to go into it all in full. Abstracts are generally a half-page paragraph or less and are more usual in longer reports.

• *Conclusions*. Sometimes, especially where analyses and interpretations are long and involved, it is helpful to highlight the main conclusions in a tail-end section so that the reader finishes with a reminder of the 'take-home' message of the investigation.

• *Appendix*. Occasionally, certain kinds of information may be incorporated into an Appendix. Such information might include the details of mathematical models or calculations, detailed background arguments, selective raw data or other aspects of the study that potentially might be of importance to readers but which would clutter up and disrupt the main report were they to be included there. Appendices are thus for informative asides that might help some readers but perhaps distract others. It follows, therefore, that appendices should be used selectively, sparingly and for a clear purpose, not as a dumping ground for odds and ends on the grounds that they might just turn out to be useful.

Example of a report

Having outlined the general principles of structuring a report, we can finish off by illustrating them more fully in a complete report. The report is one that might arise from some of the experiments we proposed earlier in the main examples, in this case aggression in crickets. As with the observational notes, the methods and results in the report are based on experiments actually carried out by students.

The effect of body size on the escalation of aggressive encounters between male field crickets (*Gryllus bimaculatus*)

Introduction

Fighting is likely to be costly in terms of time and energy expenditure and risk of injury to the individuals involved. We might thus expect natural selection to have favoured mechanisms for reducing the likelihood of costly fights. One way animals could reduce the chance of becoming involved in an escalated fight is to assess their chances of winning or losing against a given opponent before the encounter escalates into all-out fighting. There is now a substantial body of theory (e.g. Parker, 1974; Maynard Smith and Parker, 1976; Enquist *et al.*, 1985) suggesting how assessment mechanisms might evolve and much empirical evidence that animals assess each other during aggressive encounters (e.g. Davies and Halliday, 1978; Clutton-Brock *et al.*, 1979; Austad, 1983). Since the outcome of a fight is likely to be determined by some kind of difference in physical superiority between opponents, features relating to physical superiority might be expected to form the basis for assessment.

   Male field crickets (*Gryllus bimaculatus*) compete aggressively for ownership of shelters and access to females (see Simmons, 1986). Casual observation of male crickets in a sand-filled arena suggested that body size might be an important determinant of success in fights, with larger males winning more often (pers. obs.). This is borne out by Simmons (1986) who found a similar effect of body size in male *G. bimaculatus*. Observations also showed that aggressive interactions progressed through a well-defined series of escalating stages (see also Simmons, 1986) before a fight ensued. One possibility, therefore, is that these escalating stages reflect the acquisition of information about relative body size and interactions progress to the later, more aggressive, stages only when opponents are closely matched in size and the outcome is difficult to predict. This study therefore tests two predictions arising from this hypothesis:

1. large size will confer an advantage in aggressive interactions among male crickets, and

2. interactions will escalate further when opponents are more closely matched in size.

Methods

Four groups of six virgin male crickets were used in the experiment. All males were derived from separate, unrelated stock colonies a week after adult eclosion so each group comprised arbitrarily selected, unfamiliar males on establishment. Crickets were maintained on a 12 h:12 h light:dark cycle which was shifted by 4 h to allow observation at periods of peak activity (Simmons, 1986). Before establishing a group, the width of each male's pronotum (thorax) was measured at its widest point using Vernier calipers and recorded as an index of the male's body size (the pronotum was chosen because it consists of relatively inflexible cuticle that is unlikely to vary between observations or with handling; adult body size is determined at eclosion so does not change with age). The dorsal surface of the pronotum of each male was then marked with a small spot of coloured enamel paint to allow the observer to identify individuals.

Groups were established in glass arenas (60 × 60 × 30 cm) with 2-cm deep silver sand substrate. Each arena was provided with water-soaked cottonwool in a Petri dish and two to three rodent pellets. No shelters or other defendable objects were provided to avoid bias in the outcome of interactions due to positional advantages. Arenas were maintained under even 60 W white illumination in an ambient room temperature of 25 °C throughout the experiment.

Table 1 Degree of escalation increases from Aggressive stridulation to Flip. Each behaviour can thus be ascribed a rank escalation value ranging from 1 (low escalation) to 6 (high escalation)

Behaviour	Description	Escalation ranking
Aggressive stridulation	One or both males stridulate aggressively. This may occur on its own or in conjunction with other aggressive behaviours	1
Antennal lashing	One male whips his opponent with his antennae	2
Mandible spreading	One male spreads his mandibles and displays them to his opponent	3
Lunge	A male rears up and pushes forward, butting the opponent and pushing him backwards	4
Grapple	Males lock mandibles and wrestle	5
Flip	One male throws his opponent aside or onto his back. Re-engagement was rare following a Flip	6

The six males in a group were introduced into their arena simultaneously and allowed to settle for 5 min. They were then observed for 30 min during which time all encounters between males were dictated onto magnetic tape noting: (a) the individuals involved, (b) the individual initiating the encounter (the first to perform any of the components of aggressive behaviour – see below), (c) the individual that won (decided when one opponent first attempted to retreat) and (d) the components of aggressive behaviour used by each opponent during the encounter. Following Simmons (1986), the aggressive behaviours recognized here are shown in Table 1.

Results

Do larger males tend to win aggressive encounters?   To see whether larger males tended to win more often, the percentage of encounters won by each male in the four groups was plotted against pronotum width (Figure 1). A significant positive trend emerged. Figure 1, however, combined data from all four

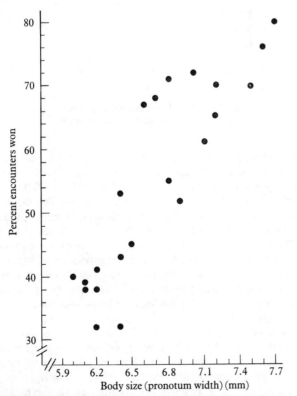

Figure 1 The relationship between the size (pronotum width) of a male and the percentage of encounters won by the male in a group of six. $r_s = 0.81$, $n = 24$, $p < 0.0001$.

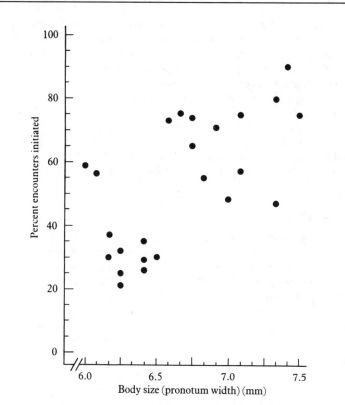

Figure 2 As Figure 1 but for the relationship between male size and the percentage number of encounters initiated by the male. $r_s = 0.63$, $n = 24$, $p < 0.03$.

groups. Did the relationship hold for each group separately? Spearman rank correlation showed a significant relationship in three of the four groups ($r_s = 0.94$, 0.99, 0.97 ($p < 0.05$ in all cases) and 0.66 (ns), $n = 6$ in all groups, one-tailed test).

If there is a size advantage as suggested by Figure 1, we might expect larger males to initiate more encounters than smaller males since they have more to gain. Figure 2 shows a significant positive correlation between pronotum width and the percentage of the recorded encounters for each male that was initiated (see Methods) by that male. As expected, therefore, larger males tended to be the initiator in more of their encounters.

One possibility that arises from Figures 1 and 2 is that the apparent effect of body size was an incidental consequence of the tendency to initiate. There may be an advantage to initiating itself, perhaps because an individual initiates only when its opponent's ability to retaliate is compromised (e.g. it is facing away from its attacker). If the males doing most of the initiating in the groups just happened to be the bigger ones, the initiation could underlie the apparent effect of body size on the chances

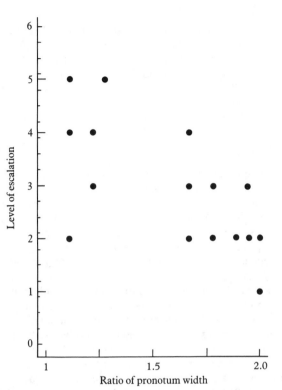

Figure 3 The relationship between ratio of pronotum widths of fighting males and the maximum level of escalation (1–6, see Methods) reached in fights. $r_s = -0.71$, $n = 20$, $p < 0.002$.

of winning. To test this, the percentage encounters won by each male when he was the initiator was compared with the percentage won when he was not. The analysis showed no significant difference ($U = 82$, $n_1, n_2 = 19$, ns*) between the two conditions.

Does difference in body size affect the degree of escalation in encounters?   Figure 3 shows the relationship between the ratio of pronotum width for pairs of opponents and the degree of escalation of their encounters. A ratio of one indicates equal size and ratios greater than one increasing departure from equality. Degree of escalation is measured as the maximum rank value (1–6, see Methods, Table 1) recorded during an encounter. As predicted, the figure shows a significant negative correlation between size ratio and degree of escalation so that escalated encounters were more likely between opponents that

* Although it is perfectly legitimate to use a Mann–Whitney U test here, the fact that we are actually comparing data for the two conditions (initiated vs. non-initiated encounters) **within** males means we could have used a different sort of two-group difference test (e.g. a Wilcoxon matched-pairs signed ranks test) which takes this into account.

were closely matched in size. The trend in Figure 3 is for data across all groups. Does the trend hold within individual males? Correlation analysis for those males (four) that were involved in five or more encounters with opponents of different relative size suggests that it did, although the trends were significant in only two cases ($r_s = -0.96$, $n = 7$, $p < 0.05$; $r_s = -0.72$, $n = 5$, ns; $r_s = -0.76$, $n = 6$, ns; $r_s = -0.99$, $n = 6$, $p < 0.05$, one-tailed test).

Discussion

The results bore out both predictions about the effects of body size on the outcome of aggressive interactions between male field crickets: larger males were more likely to win and escalation was more likely between closely matched opponents. This is consistent with the outcome of fights being largely a matter of physical superiority and with the structuring of interactions into a well-defined series of escalating stages reflecting assessment.

The fact that larger males were more likely to initiate an interaction could mean that the relative size of a potential opponent is assessible in advance of physical interaction. However, it could also reflect a general confidence effect arising from previous wins by larger males (males may initiate according to the simple decision rule 'if I won in the past, I'll probably win this time, so it is worth initiating'). Indeed, Simmons (1986) presents evidence that the number of past wins has a positive influence on the tendency for males to initiate, a result consistent with the confidence effect. Alternatively, initiation could reflect individual recognition, with males picking on those individuals against whom they have won in the past. Since this study did not record encounters independently of the performance of one of the categories of aggressive behaviour, it is not possible to say whether initiations against particular opponents occurred more or less often than expected by chance. Whatever the basis for deciding to initiate, however, there was no evidence that initiation itself conferred an advantage in terms of the outcome.

Although no resources (shelters and females) were available in the arenas, the size advantage in the aggressive interactions recorded here is in keeping with the tendency for larger males to take over shelters and mate successfully with females (Simmons, 1986). While females prefer to mate with males in or near shelters (because these provide good oviposition sites and protection from predators), they will mate with males encountered in open areas (Simmons, 1986). Aggression between males in the absence of shelters or females may thus reflect an advantage to reducing competition should a female happen to be encountered.

References

Austad, S. N. (1983) A game theoretical interpretation of male combat in the bowl and doily spider, *Frontinella pyramitela. Anim. Behav.* **19**, 59–73.

Clutton-Brock, T. H., Albon, S. D., Gibson, R. M. and Guiness, F. E. (1979) The logical stag: adaptive aspects of fighting in red deer (*Cervus elephus* L). *Anim. Behav.* **27**, 211–225.

Davies, N. B. and Halliday, T. R. (1978) Deep croaks and fighting assessment in toads, *Bufo bufo. Nature* **274**, 683–685.

Enquist, M., Plane, E. and Roed, J. (1985) Aggressive communication in fulmars (*Fulmarus glacialis*) competing for food. *Anim. Behav.* **33**, 1107–1120.

Maynard Smith, J. and Parker, G. A. (1976) The logic of asymmetric contests. *Anim. Behav.* **24**, 159–175.

Parker, G. A. (1974) Assessment strategy and the evolution of animal conflicts. *J. Theor. Biol.* **47**, 223–243.

Simmons, L. W. (1986) Inter-male competition and mating success in the field cricket, *Gryllus bimaculatus* (de Geer). *Anim. Behav.* **34**, 567–579.

SUMMARY

1. Confirmatory analyses are usually presented in summarized form (e.g. summary statistics, scattergrams) as tables or figures or in the text of a report. In all cases, sample sizes (or degrees of freedom), test statistics and probability levels should be quoted. In the case of tables and figures, these can be included within the table or figure itself or within a full, explanatory legend.

2. Results should almost never be presented as raw numerical data because these are difficult for the reader to assimilate. In the exceptional circumstances where the presentation of raw data is helpful, presentation should usually be selective to the points being made and is best incorporated as an Appendix.

3. The axes of figures should be labelled in a way that conveys their meaning clearly and succinctly. Where analyses in different figures are to be compared directly, the axes of the figures should use the same scaling.

4. The legends to tables and figures should provide a complete, self-contained explanation of what they show without the reader having to search elsewhere for relevant information.

5. Reports of investigations should be structured into clearly defined sections: Introduction, Methods, Results, Discussion,

References. Each section has a specific purpose and deals with particular kinds of information. The distinction between them should be strictly maintained. When reports are long or involved it can be helpful to add an Abstract and/or Conclusions section to highlight the main points and take-home messages.

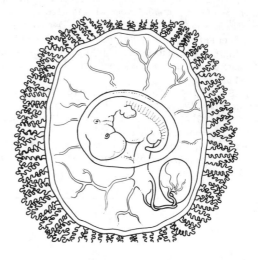

Test Finder and Help Guide

NB This key is designed to be used only with the tests described in this book.

A Looking for a difference or a trend?
   (if unsure go to *Help 1*)
      Difference   go to C
      Trend   go to B

B Trying to fit a line and/or predict new values?
   (if unsure go to *Help 2*)
      No   use Spearman rank correlation
      Yes   use linear regression

C Data measured at one level of grouping or two?
   (if unsure go to *Help 3*)
      One   go to D
      Two   go to G

D Data in **columns** of numbers
   (if unsure go to *Help 4*)
      No   go to F
      Yes   go to E

E How many columns?
      Two   use Mann–Whitney U test or
                                one-way analysis of variance
      More than two   use one-way analysis of variance

F Data are one row of counts
   (if unsure go to *Help 5*)
      Yes   use $1 \times n$ chi-squared test
      No   go to G

G Data are in both **rows** and **columns**
   (if unsure go to *Help 6*)

   Data are measurement values
   (e.g. cm, time, rank, %) or
   counts with more than one value
   per row/column cell   use two-way analysis of variance

   Data are counts with a single
   value per row/column cell   use $n \times n$ chi-squared test

HELP!

Difference or trend?

• *Difference* predictions are concerned with some kind of difference between two or more groups of data. The groups could be based on any characteristic that can be used to make a clear-cut distinction, e.g. sex, drug treatment, habitat. Thus a difference might be predicted between the growth rates of men and women, or between the development of disease in rats given drug A versus those given drug B versus those given a placebo.

• *Trend* predictions are concerned not with differences between mutually exclusive groupings but with the relationship between two more or less continuously distributed measures, e.g. the relationship between the size of a shark and the size of prey it takes, or the relationship between the amount of rainfall in a growing season and the number of apples produced by an apple tree.

Fitting a line and/or predicting a new value?

Fitting a line to a trend by linear regression involves first setting the values of x and then measuring y in relation to these values. Regression then calculates the quantitative relationship between the two measures being related rather than simply seeing whether there is some degree of association between them (which could be achieved using correlation analysis). Knowing the quantitative relationship can be important for two reasons.

1. To get an idea of the magnitude of increase or decrease in one measure with a given increase or decrease in the other, e.g. it wouldn't be surprising to find that bigger sharks ate bigger prey (a Spearman rank correlation coefficient could tell you this), but it might be important to know whether prey size increases at, say, twice the rate of shark size (only regression analysis can give you this kind of information).

2. To predict a value of one measure (e.g. the size of shark that would be dangerous to humans) from a given value of the other (e.g. prey size comparable to a human being). This example prediction could be important in deciding whether the sharks present around a beach constitute a serious threat to bathers, since it wouldn't be ethical to mix sharks and bathers of different sizes to find out by trial and error!

Levels of grouping

Many difference predictions are concerned with differences at just **one** level of grouping, e.g. differences in faecal egg counts following treatment of mice with one of four different anthelminthic drugs. Here drug treatment is the only level of grouping in which we are interested. However, if we wished, say, to distinguish between the effects of different drugs on male and female mice, we should be dealing with **two** levels of grouping: drug treatment and sex.

Columns of numbers

In these cases, each of the data groupings has more than one data value in it (not necessarily the same number of values in each case) and can be thought of as columns of values under their group headings, for example:

	Group		
	Pesticide A	Pesticide B	Control
% mortality of pest	10	30	0
	5	27	1
	3	50	1
	0	6	0
	1	3	2
	20	3	5

A single row of numbers

Here, we are dealing with a single total **count** (e.g. number of organisms, number of events, number of instances on which something met some criterion) under each of the group headings. The numbers cannot be any other kind of measure (e.g. time, length, weight, percentage, ratio). An example would be the number of people questioned who said they preferred a particular soap opera:

	Favourite soap opera			
	Soap 1	Soap 2	Soap 3	Soap 4
Number of people preferring	3	54	71	120

Rows and columns

If data have been collected at two levels of grouping, then each data value can be thought of as belonging to both a row and a column (i.e. to one row/column cell) in a table, where rows refer to one level of grouping (say sex – see Help 4) and columns to the other (drug treatment – see Help 4). If there are several values per row/column cell as below for the number of individuals dying during a period of observation:

| | | Treatment | |
		Experimental	Control
Sex	Male	3, 4, 8, 12	23, 24, 12, 32
	Female	1, 0, 2, 9	32, 45, 31, 21

then a two-way analysis of variance is appropriate. If there is just a single **count** in each cell:

| | | Treatment | |
		Experimental	Control
Sex	Male	27	91
	Female	12	129

then an $n \times n$ chi-squared test is appropriate.

SOME SELF-TEST QUESTIONS

(Answers on p. 147)

1. An experimenter recorded the following body lengths of freshwater shrimps (*Gammarus pulex*) in three different lakes.

Body size (mm)		
Lake 1	Lake 2	Lake 3
9.9	10.5	9.6
8.7	12.1	9.0
9.6	11.2	8.7
10.7	9.7	13.2
8.9	8.7	11.9
8.2	11.1	14.0
7.7	10.7	12.9
8.1	11.8	10.8

Faunal diversity in the lakes was known and the experimenter expected shrimps from more diverse lakes to be smaller because of increased interspecific competition. To test this idea he compared body lengths in each pair of lakes (1 vs. 2, 2 vs. 3 and 1 vs. 3) using Mann–Whitney U tests. Is this an appropriate analysis. If not, what would you do instead?

2. How would you decide between correlation and regression analysis when testing a trend prediction?

3. The following is part of the Discussion section of a report into the effects of temperature and weather on the reproductive rate of aphids on bean plants.

> While the results show a significant increase in the number of aphids produced as temperature rises, there is a possible confounding effect of the age of the host plant and the rate of flow of nutrients. Indeed, there was a stronger significant positive correlation between nutrient flow rate and the number of aphids produced ($r_s = 0.84, n = 20, p < 0.01$) than between temperature and production (see Results).

Do you have any criticisms of the piece?

4. What does the following tell you about the analysis from which it derives?

$$H = 14.1, \qquad \text{d.f.} = 3, \qquad p < 0.01$$

5. An agricultural researcher discovered a significant positive correlation ($r_s = 0.79$, $n = 112$, $p < 0.01$) between daily food intake and the rate of increase in body weight of pigs. What can he conclude from the correlation?

6. A plant physiologist measured the length of the third internode of some experimental plants that had received one of three different hormone treatments. The physiologist calculated the average third internode length for each treatment and for untreated control plants. The data were as follows:

	Treatment			
	Control	Hormone 1	Hormone 2	Hormone 3
Average internode length (mm)	32.3	41.6	38.4	50.2

To see whether there was any significant effect of hormone treatment, the physiologist performed a 1 × 4 chi-squared test with an expected value of 40.6 in each case and 3 degrees of freedom. Was this an appropriate test?

7. What do you understand by the terms:
(a) test statistic,
(b) ceiling effect,
(c) statistical significance?

8. Figure 1 shows a significant positive correlation, obtained in the field, between body size in female thargs and the percentage of females in each size class that were pregnant. From this, the observer concluded that male thargs preferred to mate with larger females. Is such a conclusion justified? Give reasons for your answer.

9. Why are significance tests necessary?

10. The following were the results of an experiment to look at the effect of adding an enzyme to its substrate and measuring the rate at which the substrate was split. In the control Treatment A, no enzyme was added; in Treatment B, 10 mg

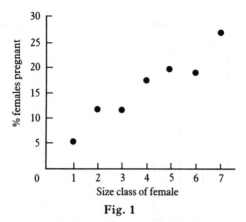

Fig. 1

of enzyme was added; in Treatment C 10 mg was added but the reaction was cooled; and in Treatment D, 10 mg was added but the reaction was warmed slightly.

Treatment A	Treatment B	Treatment C	Treatment D
0	20	52	71
1	21	69	92
2	35	100	55
1	15	32	78
0	20		105
0	24		82
			92

What predictions would you make about the outcome of the experiment and how would you analyse the data to test them?

11. A farmer called in an agricultural consultant to help him decide on the best housing conditions (those resulting in the fastest growth) for his pigs. Three types of housing were available (sty + open paddock, crating, and indoor pen). The farmer also kept four different breeds of pig and wanted to know how housing affected the growth rate of each. What analysis might the consultant perform to help the farmer reach a decision?

12. Figures 2a and 2b were used by a commercial forestry company to argue that the effect of felling on the number of bird species living in managed stands (assessed by a single standardized count in each case) was similar in both deciduous and coniferous forest. Would you agree with the company's assessment on this basis?

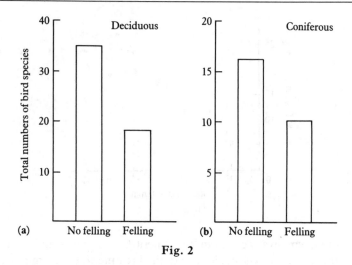

Fig. 2

13. What would a **negative** value of the test statistic r_s signify to you?

14. Derive some hypotheses and predictions from the following observational notes.

> Sampled some freshwater invertebrates from three different streams using a hand-net. There were more individuals of each species at some sites than others, both within streams and between them. Also some sites had a more or less even distribution of individuals across all species whereas others had a highly biased distribution with some species dominating the community. Some species occurred in all three streams but they tended to be smaller in some streams than others. A number of predatory dragonfly nymphs were recorded but there was never more than one species in any one sample even when more than one existed in a stream. Water quality analyses showed that one stream was badly polluted with effluent from a local factory. This stream and one of the others flowed into the third stream forming a confluence. It was noticed that stones and rocks on the substrate had fewer organisms on or under them in regions of faster flow rate.

15. Is chi-squared used for testing differences or trends?

16. A student in a hall of residence suffered from bed bugs. During the course of a week he was bitten 12 times on his legs, 3 times on his torso, 6 times on his arms and once on his head. Could these data be analysed for site preferences by the bugs? If so, how?

17. An experimenter had counted the number of times kittens showed elements of play behaviour when they were in the

presence of their mother or their father and with or without a same-sex sibling. The experimenter had collected ten counts for each condition: (a) mother/sibling, (b) mother/no sibling, (c) father/sibling, (d) father/no sibling, and was trying to decide between a 2×2 chi-squared analysis and a 2×2 two-way analysis of variance. What would you suggest and why?

18. Why do biologists regard a probability of 5% or less as the criterion for significance? Why not be even stricter and use 1%?

19. A fisheries biologist was interested in the maximum size of prey that was acceptable to adult barracuda. To find out what it was, he introduced six adult barracuda into separate tanks and fed them successively larger species of fish (all known to coexist with barracuda in the wild). He then calculated the mean size of the fish that the barracuda last accepted before refusing a fish as a measure of the maximum size they would take. Is this a sensible procedure?

20. A psychologist argued that since males of a species of monkey had larger brains than the females there was less point in trying to teach females complex problem-solving exercises. Any comments?

APPENDIX I

Table of nonparametric confidence limits
to the median

Sample size (n)	r (for p approx. 95%)
2	–
3	–
4	–
5	–
6	1
7	1
8	1
9	2
10	2
11	2
12	3
13	3
14	3
15	4
16	4
17	5
18	5
19	5
20	6
21	6
22	6
23	7
24	7
25	8
26	8
27	8
28	9
29	9
30	10

r denotes the number of values in from the
extremes of the data set that identifies the
95% confidence limits (see text). Modified
after Colquhoun (1971) *Lectures on bio-statistics,* Clarendon Press, Oxford.

APPENDIX II

A. EXAMPLE OF A MANN–WHITNEY TEST

An ecologist is interested in the effect of microhabitat on the distribution of periwinkles on a rocky shore. Two habitats – a boulder/shingle beach and crevices in a rocky stack – were compared for the prevalence of the commonest periwinkle species measured as the percentage of the total number of all individuals of invertebrate species recorded within quadrat samples. A Mann–Whitney U test comparing the percentages in replicates of the two habitats is as follows:

% in boulder/shingle	Rank	% in crevices	Rank
40	3.5	60	11
37	1	40	3.5
41	5	55	9
39	2	57	10
43	6	51	8
		63	12
		49	7
	$R_1 = 17.5$		$R_2 = 60.5$

$$U_1 = n_1 \cdot n_2 + ((n_1(n_1 + 1))/2) - R_1$$

$$n_1 = 5, \; n_2 = 7$$

$$U_1 = 5 \times 7 + (5 \times 6)/2 - 17.5$$

$$= 35 + 15 - 17.5 = 32.5$$

$$n_1 \times n_2 - U_1 = 5 \times 7 - 32.5 = 2.5$$

Since 2.5 is less than 32.5, $U = 2.5$.

Looking this up in Appendix III, Table B, we see that U is smaller than the critical value of 7 for $p < 0.05$, so we can conclude

that there is a significant difference between habitat types in the prevalence of the periwinkle species.

B. EXAMPLE OF A NONPARAMETRIC ONE-WAY ANALYSIS OF VARIANCE
(modified after Meddis, 1984)

An experimenter measured how much alcohol students in all-female, all-male or mixed halls of residence drank in a month. There are two types of prediction that can be tested:

1. that types of hall differ in their alcohol consumption (general prediction);
2. that all-male halls get through more alcohol than mixed halls and that all-female halls drink the least (specific prediction).

| | | | Groups ($i = 3$) Units of alcohol drunk | | | |
| | 1 = female | | 2 = mixed | | 3 = male | |
	Units	Rank	Units	Rank	Units	Rank
	38	1.5	38	1.5	80	9.5
	47	3.5	60	6	80	9.5
	47	3.5	61	7	85	12.5
	58	5	81	11	87	14
	65	8	85	12.5	91	15
$n_i =$	5		5		5	
$R_i =$		21.5		38		60.5
Mean rank		4.3		7.6		12.1
$N = \sum n_i = 5 + 5 + 5 = 15$						

The data have been ranked in the table, and then the sums of the ranks for each group have been calculated.

General prediction: Students in the different types of hall differ in their alcohol consumption?

Compute

$$H = \frac{12}{N(N + 1)} (\Sigma R_i^2 / n_i) - 3(N + 1)$$

$$H = (12/(15 \times 16)) \cdot (21.5^2/5 + 38^2/5 + 60.5^2/5) - 3(16)$$

$$= 7.665$$

for 2 degrees of freedom; the critical value for H (as χ^2) is 5.99 (Appendix III, Table A). Therefore, this value of H is significant at the 5% level.

Specific prediction: The levels of alcohol consumption follow the order all-female halls < mixed halls < all-male halls?

The hypothesis is that female < mixed < male, hence the values of λ_i are 1, 2, and 3, since this is the specified rank order of the mean ranks.

Thus

$$L = \Sigma\lambda_i R_i = (1)(21.5) + (2)(38) + (3)(60.5) = 279$$
$$\Sigma\lambda_i n_i = (1)(5) + (2)(5) + (3)(5) = 30$$
$$\Sigma n_i(\lambda_i)^2 = (1)(1)(5) + (2)(2)(5) + (3)(3)(5) = 70$$

then

$$E = (N + 1)[(\Sigma\lambda_i n_i)/2] = 16(30/2) = 240$$

and

$$V = (N + 1)[N \cdot \Sigma n_i \lambda_i^2 - (\Sigma\lambda_i n_i)^2]/12$$
$$= 16(15(70) - 30^2)/12 = 200$$

$$z = (L - E)/\sqrt{V} = (279 - 240)/\sqrt{200} = 2.76$$

Since this is greater than 1.64 (see Appendix III, Table C), the specified order of mean ranks is significant at the 5% level: therefore $p < 0.05$.

C.   EXAMPLE OF A $1 \times N$ CHI-SQUARED TEST

A clinical microbiologist was assaying the effect of four antibiotics on a bacterial culture. To see whether the antibiotics differed in their ability to kill the bacterium, he counted the number of cultures on which clear plaques appeared after drop treatment with each antibiotic and performed a 1×4 chi-squared test assuming equal expected values across antibiotics. The results were as follows:

	Treatment			
	Antibiotic A	Antibiotic B	Antibiotic C	Antibiotic D
Observed no. with plaques	21	7	18	30
Expected no.	19	19	19	19

$$\chi^2 = \Sigma \frac{(O - E)^2}{E}$$

$$= \frac{(21 - 19)^2}{19} + \frac{(7 - 19)^2}{19} + \frac{(18 - 19)^2}{19} + \frac{(30 - 19)^2}{19}$$

$$= 0.21 + 7.58 + 0.05 + 6.37$$

$$= 14.21$$

degrees of freedom $= 4 - 1 = 3$

Checking $\chi^2 = 14.21$ for 3 d.f. in Appendix III, Table A shows that it is significant at $p < 0.01$.

D.   EXAMPLE OF A NONPARAMETRIC TWO-WAY ANALYSIS OF VARIANCE
(modified after Meddis, 1984)

A team of researchers interested in the consequences of regular running for various measures of health and well-being investigated the joint effects of fat in the diet and running for three or more days a week on levels of blood cholesterol. They measured cholesterol in people who had low or high fat intake and did or did not run on three or more days a week. They then cast their cholesterol measures as follows and carried out a nonparametric 'two-way analysis of variance'.

| | Cholesterol scores | |
	Did not run	Did run
High fat diet	23, 24, 25, 26, 28	18, 18.5, 19.1, 20.1, 20.5
	= Group A	= Group B
Low fat diet	19, 20, 20, 21, 23	3, 4, 5, 6, 7
	= Group C	= Group D

Case A: specific tests

1. First we shall assume the team wanted to test some specific predictions (Pr):

    Pr1: *Running will reduce blood cholesterol (i.e. $B + D < A + C$ in the table above).*

Pr2: *Low fat diet will reduce blood cholesterol (i.e.* $C + D < A + B$).

Pr3: *The effect of running will be greater when on a low fat diet (i.e.* $C - D > A - B$).

2. Now we rank all the data values irrespective of cell in the table, giving low ranks to low values and averaging ranks for tied values. Thus we arrive at the following table:

	Rank values	
	Did not run	Did run
High fat diet	15.5, 17, 18, 19, 20	6, 7, 9, 12, 13
	= Group A	= Group B
Low fat diet	8, 10.5, 10.5, 14, 15.5	1, 2, 3, 4, 5
	= Group C	= Group D

3. Work out the sums of the ranks and sample sizes for each cell. These are:

	$i = 1$	$i = 2$	$i = 3$	$i = 4$	Sum
Group (i)	A	B	C	D	
Rank sums (R_i)	89.5	47	58.5	15	210
Sample size (n_i)	5	5	5	5	20
Mean ranks	17.9	9.4	11.7	3	

Check that the ranking is correct by using N, the total sample size, to calculate $N(N + 1)/2$. This should equal the sum of ranks. Here $N(N + 1)/2 = 20(21)/2 = 210$, so the ranking is alright.

4. Test the predictions by obtaining the coefficients, λ_i, for each prediction and then calculating the test statistic z using:

$$z = (L - E)/\sqrt{V}$$

where

$$L = \Sigma \lambda_i R_i$$
$$E = (N + 1)\Sigma n_i \lambda_i / 2$$
$$V = (N + 1)[N \Sigma n_i \lambda_i^2 - (\Sigma n_i \lambda_i)^2]/12$$

Thus for Pr1, the prediction is that $B + D < A + C$, i.e. $+A - B + C - D > 0$, so the coefficients, λ_i, are: $+1, -1, +1, -1$.

$$\Sigma n_i \lambda_i = +5 - 5 + 5 - 5, \qquad \text{sum} = 0$$

$$\Sigma n_i \lambda_i^2 = +5 + 5 + 5 + 5, \qquad \text{sum} = 20$$

Thus:

$$L = \Sigma \lambda_i R_i$$

$$= (+1)(89.5) + (-1)(47) + (+1)(58.5) + (-1)(15)$$

$$= 89.5 - 47 + 58.5 - 15$$

$$= 86$$

$$E = (N + 1)(\Sigma n_i \lambda_i)/2$$

$$= (21)(0)/2$$

$$= 0$$

$$V = (n + 1)[\sum n_i \lambda_i^2 - (\sum n_i \lambda_i)^2]/12$$

$$= (21)[20(20) - (0)^2]/12$$

$$= 700$$

$$\therefore \; z = (86 - 0)/\sqrt{700}$$

$$= 3.25$$

Looking z up in Appendix III, Table C we find that a value of 3.25 means there is less than 0.1% ($p < 0.001$) probability that we could have obtained our predicted order by chance. We therefore accept that the fit to our predicted order is significant.

Prediction Pr2 is that $C + D < A + B$, i.e. that $+A + B - C - D > 0$, with coefficients $+1, +1, -1, -1$ and

$$\Sigma n_i \lambda_i = +5 + 5 - 5 - 5, \qquad \text{sum} = 0$$

$$\Sigma n_i \lambda_i^2 = +5 + 5 + 5 + 5, \qquad \text{sum} = 20$$

Thus:

$$L = (+1)(89.5) + (+1)(47) + (-1)(58.5) + (-1)(15)$$

$$= 89.5 + 47 - 58.5 - 15$$

$$= 63$$

$$E = (21)(0)/2$$

$$= 0$$

$$V = (21)[(20(20) - (0)^2]/12$$

$$= 700$$

$$\therefore \ z = (63 - 0)/\sqrt{700}$$

$$= 2.38$$

Looking in Table C, Appendix III, again, we see that $z = 2.38$ has an associated probability of occurring by chance of $p < 0.01$. Again, therefore, the fit to our prediction is significant.

For Pr3, $C - D > A - B$, i.e. $-A + B + C - D > 0$, so λ_i are $-1, +1, +1, -1$.

$$\Sigma n_i \lambda_i = -5 + 5 + 5 - 5, \qquad \text{sum} = 0$$

$$\Sigma n_i \lambda_i^2 = +5 + 5 + 5 + 5, \qquad \text{sum} = 20$$

$$L = -89.5 + 47 + 58.5 - 15$$

$$= 1$$

$$E = 21(0)/2$$

$$= 0$$

$$V = (21)[20(20) - (0)^2]/12$$

$$= 700$$

$$z = (1 - 0)/\sqrt{700}$$

$$= 0.04$$

This time the associated probability for z is not significant ($p > 0.05$).

Thus the three specific predictions have been tested and the team can conclude that:

1. Running is associated with a reduction in blood cholesterol ($z = 3.25$, $p < 0.001$).

2. Low fat diet is associated with low blood cholesterol ($z = 2.38$, $p < 0.01$).

3. The effect of running is not greater when on a low fat diet ($z = 0.04$, ns).

Case B: general tests

Instead of making specific predictions, the team could have made some general ones. However, the only time it makes sense to do **both** is after **none** of the specific predictions have been supported by the data. If none of the predicted patterns emerges it is then useful to ask whether there are any differences at all between groups.

General predictions can be tested **only** when there are equal numbers of measures in each cell of the analysis of variance table.

1. The team first formulate their general predictions:

   Pr4: *Running will affect blood cholesterol levels.*

   Pr5: *The amount of fat in the diet will affect blood cholesterol levels.*

   Pr6: *There will be an interaction between running and diet in their effects on blood cholesterol levels.*

2. Testing the predictions:

   Pr4: Sum the ranks for exercise groups as A + C and B + D giving:

$$R_1 = 89.5 + 58.5$$
$$= 148, n = 10$$
$$R_2 = 47 + 15$$
$$= 62, n = 10$$

Then

$$H_1 = \left[\frac{12}{N(N+1)} \Sigma \left(\frac{R_i^2}{n_i} \right) - 3(N+1) \right]$$

$$= \left[\frac{12}{20(21)} \left(\frac{148^2}{10} + \frac{62^2}{10} \right) - 3(21) \right]$$

$$= 10.57$$

Looking up H as χ^2 in Appendix III, Table A shows $H = 10.57$, d.f. $= 1$ to be significant at $p < 0.01$.

Pr5: Sum ranks for A + B and C + D, giving:

$$R_1 = 89.5 + 47$$
$$= 136.5, n = 10$$
$$R_2 = 58.5 + 15$$
$$= 73.5, n = 10$$

$$H_2 = \left[\frac{12}{20(21)} \left(\frac{136.5^2}{10} + \frac{73.5^2}{10} \right) - 3(21) \right]$$

$$= 5.67$$

Table A shows $H = 5.67$, d.f. $= 1$ to have an associated probability of $p < 0.05$.

Pr6: First calculate H for the total number of cells using the four rank sums:

$$H_{tot} = \left[\frac{12}{20(21)} \left(\frac{89.5^2}{5} + \frac{47^2}{5} + \frac{58.5^2}{5} + \frac{15^2}{5} \right) - 3(21) \right]$$

$$= 16.24$$

Then

$$H_3 = H_{tot} - H_1 - H_2$$

$$= 16.24 - 10.57 - 5.67$$

$$= 0$$

Table A shows $H = 0$, d.f. $= 1$ to have an associated probability of $p > 0.05$.

The team can thus conclude that:

1. Running **does** affect blood cholesterol levels ($H = 10.57$, $p < 0.01$).

2. The amount of fat in the diet **does** influence blood cholesterol levels ($H = 5.67$, $p < 0.05$).

3. There is no interaction between running and fat in the diet in determining blood cholesterol levels.

E.   EXAMPLE OF A SPEARMAN RANK CORRELATION ANALYSIS

The following data set shows the number of sexual approaches by female chaffinches (y) to males with different plumage brightness scores (x). Brightness scores were compounds of different criteria totalling to a maximum possible value of 60. Is there any correlation between male brightness and sexual approaches?

Male	Brightness score	No. approaches by females
1	17	0
2	25	3
3	22	2
4	43	6
5	30	5
6	56	8
7	31	4
8	47	7
9	29	2
10	31	4

Ranking the scores for x and y and calculating the difference (d_i) and square of the difference (d_i^2) between ranks gives:

Rank brightness	Rank no. approaches	d_i	d_i^2
1	1	0	0
3	4	1	1
2	2.5	0.5	0.25
8	8	0	0
5	7	2	4
10	10	0	0
6.5	5.5	1	1
9	9	0	0
4	2.5	1.5	2.25
6.5	5.5	1	1
			$\sum d_i^2 = 9.5$

The Spearman rank correlation coefficient, r_s, can now be calculated as:

$$r_s = 1 - [(6\Sigma d_i^2)/(n^3 - n)]$$
$$= 1 - [6(9.5)/(1000 - 10)]$$
$$= 1 - 0.058$$
$$= 0.942$$

Reference to Appendix III, Table D shows that for $n = 10$, this value is significant at $p < 0.01$ (two-tailed). We can therefore conclude that approaches are associated with male brightness.

F.   EXAMPLE OF A LINEAR REGRESSION ANALYSIS

A biogeographer was interested in the effect of distance from one of her study sites on the number of new species discovered on similar sites. Assuming that her site acted as a source region for the others, she predicted that, as distance increased, the number of new species would also increase. She chose five sites at 10-km intervals (x-values) and performed the following linear regression analysis:

	Distance from site (km)	No. of new species
	10	22
	20	23
	30	25
	40	27
	50	28
Sum	150	125

1. Calculate x^2 and y^2 and the product xy for every pair.

	x^2	y^2	xy
	100	484	220
	400	529	460
	900	625	750
	1600	729	1080
	2500	784	1400
Sum	5500	3151	3910

2. Calculate

$$S_{xx} = \Sigma x^2 - \frac{(\Sigma x)^2}{n}$$

$$= 5500 - \frac{150^2}{5}$$

$$= 1000$$

3. Calculate

$$S_{yy} = \Sigma y^2 - \frac{(\Sigma y)^2}{n}$$

$$= 3151 - \frac{125^2}{5}$$

$$= 26$$

4. Calculate

$$S_{xy} = \Sigma xy - \frac{(\Sigma x)(\Sigma y)}{n}$$

$$= 3910 - \frac{(150)(125)}{5}$$

$$= 160$$

5. Calculate the slope of the regression line as:

$$b = S_{xy}/S_{xx}$$

$$= 160/1000$$

$$= 0.16$$

6. Calculate the intercept of the line as:

$$a = \bar{y} - b\bar{x}$$

$$= (125/5) - 0.16(150/5)$$

$$= 20.2$$

7. Calculate the standard error to the slope as:

$$s_{y/x}^2 = (1/(n-2))(S_{yy} - S_{xy}^2/S_{xx})$$

$$= (1/(5-2))(26 - 160^2/1000)$$

$$= 0.1333$$

$$\text{s.e.} = \sqrt{(s_{y/x}^2/S_{xx})}$$

$$= \sqrt{(0.1333/1000)}$$

$$= 0.011$$

8. Calculate the test statistic F as follows:

$$\text{Regression sum of squares (RSS)} = (S_{xy})^2/S_{xx}$$

$$= 160^2/1000$$

$$= 25.6$$

$$\text{Regression mean square (RMS)} = \text{RSS}/1$$

$$= 25.6$$

$$\text{Deviation sum of squares (DSS)} = S_{yy} - (S_{xy})^2/S_{xx}$$

$$= 26 - 160^2/1000$$

$$= 0.4$$

$$\text{Deviation mean square (DMS)} = \text{DSS}/(n-2)$$

$$= 0.4/(5-2)$$

$$= 0.1333$$

$$F = \text{RMS}/\text{DMS}$$

$$= 25.6/0.1333$$

$$= 192.0$$

$F = 192$ can now be checked against Appendix III, Table F for 1 (f_1) and $n - 2 = 3$ (f_2) degrees of freedom, yielding a probability of $p < 0.01$. Our biogeographer can thus conclude that there is a significant positive relationship between distance from her original site and the number of new species.

9. Predicting a new y value and testing an observed y for departure from prediction.

Having established that there was a general relationship between distance and number of species, the biogeographer wanted to check its predictive value by seeing how closely it could predict the number of new species at a site not included in the original analysis. She therefore chose a new site at 25 km from the original site.

To predict y for her new x-value (x'), she used the linear regression equation $y = a + bx$ incorporating the desired x'-value of 25. The predicted y-value from the equation was thus:

$$y = 20.2 + 0.16(25)$$

$$= 24.2$$

The biogeographer then sampled a site at the new distance and came up with 23 new species, the same as at the 20-km site. To see whether this observed number differed significantly from the prediction of 24.2, the biogeographer calculated the test statistic t as:

$$t = \frac{\text{observed } y - \text{predicted } y}{\text{standard error of the prediction}}$$

where the standard error is calculated as:

$$\text{s.e.} = \sqrt{((s_{y/x}^2)(1 + 1/n + (x' - \bar{x})/S_{xx}))}$$

$$= \sqrt{((0.1333)(1 + 1/5 + (25 - 30)^2/1000))}$$

$$= \sqrt{0.1633}$$

$$= 0.4041$$

Calculating t for observed and predicted y-values:

$$t = \frac{23 - 24.2}{0.4008}$$

$$= 2.99$$

$$\text{Degrees of freedom} = n - 2$$

$$= 5 - 2$$

$$= 3$$

Checking in Appendix III, Table E shows that $t = 2.99$ for d.f. $= 3$ has an associated probability of $p > 0.05$. Therefore there is no significant difference between the biogeographer's observed y-value and that predicted by the regression equation.

APPENDIX III

Table A Critical values of chi-squared at different levels of p. To be significant, calculated values must be **greater** than those in the table for the chosen level (0.05, 0.01, 0.001) of p and the appropriate number of degrees of freedom

Degrees of freedom	Probability level		
	0.05	0.01	0.001
1	3.841	6.635	10.83
2	5.991	9.210	13.82
3	7.815	11.34	16.27
4	9.488	13.28	18.47
5	11.07	15.09	20.51
6	12.59	16.81	22.46
7	14.07	18.48	24.32
8	15.51	20.09	26.13
9	16.92	21.67	27.88
10	18.31	23.21	29.59

Table B Critical values of Mann–Whitney U at $p = 0.05$. To be significant, values must be **smaller** than those in the table for appropriate sizes of n_1 and n_2

n_1 \ n_2	3	4	5	6	7	8	9	10	15	20
2	–	–	0	0	0	0	0	0	1	2
3	0	0	1	2	2	2	2	3	5	8
4		1	2	3	4	4	4	5	10	13
5			4	5	6	7	7	8	14	20
6				7	8	10	10	11	19	27
7					11	12	12	14	24	34
8						15	15	17	29	41
9							17	20	34	48
10							20	23	39	55
15							34	39	64	90
20							48	55	90	127

Table C Probabilities associated with different values of z. The body of the table shows probabilities associated with different values of z. Values of z given to the first decimal place vertically and the second decimal place horizontally. z must therefore exceed 1.64 to be significant at $p < 0.05$

z	.00	.01	.02	.03	.04	.05	.06	.07	.08	.09
1.5	.0668	.0655	.0643	.0630	.0618	.0606	.0594	.0582	.0571	.0559
1.6	.0548	.0537	.0526	.0516	.0505	.0495	.0485	.0475	.0465	.0455
1.7	.0446	.0436	.0427	.0418	.0409	.0401	.0392	.0384	.0375	.0367
1.8	.0359	.0351	.0344	.0336	.0329	.0322	.0314	.0307	.0301	.0294
1.9	.0287	.0281	.0274	.0268	.0262	.0256	.0250	.0244	.0239	.0233
2.0	.0228	.0222	.0217	.0212	.0207	.0202	.0197	.0192	.0188	.0183
2.1	.0179	.0174	.0170	.0166	.0162	.0158	.0154	.0150	.0146	.0143
2.2	.0139	.0136	.0132	.0129	.0125	.0122	.0119	.0116	.0113	.0110
2.3	.0107	.0104	.0102	.0099	.0096	.0094	.0091	.0089	.0087	.0084
2.4	.0082	.0080	.0078	.0075	.0073	.0071	.0069	.0068	.0066	.0064
2.5	.0062	.0060	.0059	.0057	.0055	.0054	.0052	.0051	.0049	.0048
2.6	.0047	.0045	.0044	.0043	.0041	.0040	.0039	.0038	.0037	.0036
2.7	.0035	.0034	.0033	.0032	.0031	.0030	.0029	.0028	.0027	.0026
2.8	.0026	.0025	.0024	.0023	.0023	.0022	.0021	.0021	.0020	.0019
2.9	.0019	.0018	.0018	.0017	.0016	.0016	.0015	.0015	.0014	.0014
3.0	.0013	.0013	.0013	.0012	.0012	.0011	.0011	.0011	.0010	.0010
3.1	.0010	.0009	.0009	.0009	.0008	.0008	.0008	.0008	.0007	.0007
3.2	.0007									
3.3	.0005									
3.4	.0003									
3.5	.00023									
3.6	.00016									
3.7	.00011									
3.8	.00007									
3.9	.00005									
4.0	.00003									

Table D Critical values for the Spearman rank correlation coefficient r_s. Values must be **greater** than those in the table to be significant at the indicated level of probability

N	prob. 0.05 prob. 0.10	0.025 0.05	0.01 0.02	0.005 0.01	(one-tailed) (two-tailed)
4	1.000				
5	.900	1.000	1.000		
6	.829	.886	.943	1.000	
7	.714	.786	.893	.929	
8	.643	.738	.833	.881	
9	.600	.700	.783	.833	
10	.564	.648	.745	.794	
11	.536	.618	.709	.755	
12	.503	.587	.671	.726	
13	.484	.560	.648	.703	
14	.464	.538	.622	.675	
15	.443	.521	.604	.654	
16	.429	.503	.582	.635	
17	.414	.485	.566	.615	
18	.401	.472	.550	.600	
19	.391	.460	.535	.584	
20	.380	.447	.520	.570	
21	.370	.435	.508	.556	
22	.361	.425	.496	.544	
23	.353	.415	.486	.532	
24	.344	.406	.476	.521	
25	.337	.398	.466	.511	

Table E Critical values of t at different levels of p. To be significant at the appropriate level of probability, values must be **greater** than those in the table for the appropriate degrees of freedom

Degrees of freedom	Level of probability					
	0.10	0.05	0.02	0.01	0.002	0.001
1	6.314	12.71	31.82	63.66	318.3	636.6
2	2.920	4.303	6.965	9.925	22.33	31.60
3	2.353	3.182	4.541	5.841	10.21	12.92
4	2.132	2.776	3.747	4.604	7.173	8.610
5	2.015	2.571	3.365	4.032	5.893	6.869
6	1.942	2.447	3.143	3.707	5.208	5.959
7	1.895	2.365	2.998	3.499	4.785	5.408
8	1.860	2.306	2.896	3.355	4.501	5.041
9	1.833	2.262	2.821	3.250	4.297	4.781
10	1.812	2.228	2.764	3.169	4.144	4.587
11	1.796	2.201	2.718	3.106	4.025	4.437
12	1.782	2.179	2.681	3.055	3.930	4.318
13	1.771	2.160	2.650	3.012	3.852	4.221
14	1.761	2.145	2.624	2.977	3.787	4.140
15	1.753	2.131	2.602	2.947	3.733	4.073
16	1.746	2.120	2.583	2.921	3.686	4.015
17	1.740	2.110	2.567	2.898	3.646	3.965
18	1.734	2.101	2.552	2.878	3.610	3.922
19	1.729	2.093	2.539	2.861	3.579	3.883
20	1.725	2.086	2.528	2.845	3.552	3.850

Table F Critical values of F at $p = 0.05$ (upper) and $p = 0.01$ (P146). To be significant, values must be **greater** than those in the tables for the appropriate degrees of freedom (f_1 and f_2)

$p = 0.05$

f_2 \ f_1	1	2	3	4	5	6	7	8	9	10	12	15	20	30	∞
1	161.4	199.5	215.7	224.6	230.2	234.0	236.8	238.9	240.5	241.9	243.9	245.9	248.0	250.1	254.3
2	18.51	19.00	19.16	19.25	19.30	19.33	19.35	19.37	19.38	19.40	19.41	19.43	19.45	19.46	19.50
3	10.13	9.55	9.28	9.12	9.01	8.94	8.89	8.85	8.81	8.79	8.74	8.70	8.66	8.62	8.53
4	7.71	6.94	6.59	6.39	6.26	6.16	6.09	6.04	6.00	5.96	5.91	5.86	5.80	5.75	5.63
5	6.61	5.79	5.41	5.19	5.05	4.95	4.88	4.82	4.77	4.74	4.68	4.62	4.56	4.50	4.36
6	5.99	5.14	4.76	4.53	4.39	4.28	4.21	4.15	4.10	4.06	4.00	3.94	3.87	3.81	3.67
7	5.59	4.74	4.35	4.12	3.97	3.87	3.79	3.73	3.68	3.64	3.57	3.51	3.44	3.38	3.23
8	5.32	4.46	4.07	3.84	3.69	3.58	3.50	3.44	3.39	3.35	3.28	3.22	3.15	3.08	2.93
9	5.12	4.26	3.86	3.63	3.48	3.37	3.29	3.23	3.18	3.14	3.07	3.01	2.94	2.86	2.71
10	4.96	4.10	3.71	3.48	3.33	3.22	3.14	3.07	3.02	2.98	2.91	2.85	2.77	2.70	2.54
11	4.84	3.98	3.59	3.36	3.20	3.09	3.01	2.95	2.90	2.85	2.79	2.72	2.65	2.57	2.40
12	4.75	3.89	3.49	3.26	3.11	3.00	2.91	2.85	2.80	2.75	2.69	2.62	2.54	2.47	2.30
13	4.67	3.81	3.41	3.18	3.03	2.92	2.83	2.77	2.71	2.67	2.60	2.53	2.46	2.38	2.21
14	4.60	3.74	3.34	3.11	2.96	2.85	2.76	2.70	2.65	2.60	2.53	2.46	2.39	2.31	2.13
15	4.54	3.68	3.29	3.06	2.90	2.79	2.71	2.64	2.59	2.54	2.48	2.40	2.33	2.25	2.07
16	4.49	3.63	3.24	3.01	2.85	2.74	2.66	2.59	2.54	2.49	2.42	2.35	2.28	2.19	2.01
17	4.45	3.59	3.20	2.96	2.81	2.70	2.61	2.55	2.49	2.45	2.38	2.31	2.23	2.15	1.96
18	4.41	3.55	3.16	2.93	2.77	2.66	2.58	2.51	2.46	2.41	2.34	2.27	2.19	2.11	1.92
19	4.38	3.52	3.13	2.90	2.74	2.63	2.54	2.48	2.42	2.38	2.31	2.23	2.16	2.07	1.88
20	4.35	3.49	3.10	2.87	2.71	2.60	2.51	2.45	2.39	2.35	2.28	2.20	2.12	2.04	1.84

$p = 0.01$

f_1 / f_2	1	2	3	4	5	6	7	8	9	10	12	15	20	30	∞
1	4052	4999	5403	5625	5764	5859	5928	5982	6022	6106	6157	6209	6157	6261	6366
2	98.50	99.00	99.17	99.25	99.30	99.33	99.36	99.37	99.39	99.40	99.42	99.43	99.45	99.47	99.50
3	34.12	30.82	29.46	28.71	28.24	27.91	27.67	27.49	27.35	27.23	27.05	26.87	26.69	26.50	26.13
4	21.20	18.00	16.69	15.98	15.52	15.21	14.98	14.80	14.66	14.55	14.37	14.20	14.02	13.84	13.46
5	16.26	13.27	12.06	11.39	10.97	10.67	10.46	10.29	10.16	10.05	9.89	9.72	9.55	9.38	9.02
6	13.75	10.92	9.78	9.15	8.75	8.47	8.26	8.10	7.98	7.87	7.72	7.56	7.40	7.23	6.88
7	12.25	9.55	8.45	7.85	7.46	7.19	6.99	6.84	6.72	6.62	6.47	6.31	6.16	5.99	5.65
8	11.26	8.65	7.59	7.01	6.63	6.37	6.18	6.03	5.91	5.81	5.67	5.52	5.36	5.20	4.86
9	10.56	8.02	6.99	6.42	6.06	5.80	5.61	5.47	5.35	5.26	5.11	4.96	4.81	4.65	4.31
10	10.04	7.56	6.55	5.99	5.64	5.39	5.20	5.06	4.94	4.85	4.71	4.56	4.41	4.25	3.91
11	9.65	7.21	6.22	5.67	5.32	5.07	4.89	4.74	4.63	4.54	4.40	4.25	4.10	3.94	3.60
12	9.33	6.93	5.95	5.41	5.06	4.82	4.64	4.50	4.39	4.30	4.16	4.01	3.86	3.70	3.36
13	9.07	6.70	5.74	5.21	4.86	4.62	4.44	4.30	4.19	4.10	3.96	3.82	3.66	3.51	3.17
14	8.86	6.51	5.56	5.04	4.69	4.46	4.28	4.14	4.03	3.94	3.80	3.66	3.51	3.35	3.00
15	8.68	6.36	5.42	4.89	4.56	4.32	4.14	4.00	3.89	3.80	3.67	3.52	3.37	3.21	2.87
16	8.53	6.23	5.29	4.77	4.44	4.20	4.03	3.89	3.78	3.69	3.55	3.41	3.26	3.10	2.75
17	8.40	6.11	5.18	4.67	4.34	4.10	3.93	3.79	3.68	3.59	3.46	3.31	3.16	3.00	2.65
18	8.29	6.01	5.09	4.58	4.25	4.01	3.84	3.71	3.60	3.51	3.37	3.23	3.08	2.92	2.57
19	8.18	5.93	5.01	4.50	4.17	3.94	3.77	3.63	3.52	3.43	3.30	3.15	3.00	2.84	2.49
20	8.10	5.85	4.94	4.43	4.10	3.87	3.70	3.56	3.46	3.37	3.23	3.09	2.94	2.78	2.42

Answers to Self-test Questions

1. No, this is not an appropriate analysis because it involves the multiple use of a two-group test (Mann–Whitney U test). A significant result becomes more likely by chance the greater the number of two-group comparisons that are made. An appropriate analysis would be a nonparametric one-way analysis of variance which allows a test of the specific prediction that body size will be greatest in Lake 1, intermediate in Lake 2 and least in Lake 3.

2. Regression analysis is appropriate when testing for cause and effect in the relationship between x- and y-values, and the x-values are established in the experiment. The data are measured on some kind of constant interval scale that allows a precise, quantitative relationship to be calculated. Because it depends on establishing a quantitative relationship, predicting new values of one variable from new values of the other also demands regression analysis. In other cases, correlation analysis is necessary. Nonparametric correlation analysis is appropriate for both these kinds of data, and others where x-values are merely measured, but yields only the sign and magnitude of the relationship and makes no assumptions about the cause-and-effect relationship between x- and y-values.

3. The investigator has introduced a new analysis into the Discussion (the correlation between nutrient flow rate and aphid production). The analysis should, of course, be in the Results section.

4. The information tells you that the investigator carried out a nonparametric one-way analysis of variance and that he/she tested a general prediction, thus using the test statistic H instead of z. The 3 degrees of freedom tells you that the analysis compared four groups and the p-value that $H = 14.1$ at d.f. $= 3$ is significant at the 1% level.

5. He can conclude that there is a significant positive association between daily food intake and growth rate, but he can't necessarily infer that increased growth rate is caused by greater food intake; it could be that faster growing pigs simply eat more food as a result.

6. No, a chi-squared test is not appropriate here because the data are constant interval measurements and not counts.

7(a). A test statistic is calculated by a significance test and its value has a known probability of occurring by chance for any given sample size or number of degrees of freedom.

(b). A ceiling effect occurs where observational or experimental procedures are too undemanding to allow a prediction to be tested; all samples approach the maximum value.

(c). Statistical significance refers to cases where the probability that a difference or trend as extreme as the one observed could have occurred by chance, if the null hypothesis of no difference or trend is true, is equal to or less than an accepted threshold probability (usually 5%, but sometimes 1% or 10%).

8. No, the observer cannot conclude that male thargs prefer larger females just from this. It could be, for instance, that larger females are simply more mature and thus more likely to conceive. Alternatively, depending on how and when size was measured, pregnant females may be larger precisely because they are pregnant!

9. Significance tests provide a generally accepted, arbitrary yardstick for deciding whether a difference or trend is interesting. The yardstick is the probability that the observed difference or trend could have occurred by chance when there wasn't really such a difference or trend in the population. Random variation in sampling will mean that differences or trends will crop up from time to time just by chance.

10. A reasonable prediction would be that the rate of reaction would be lowest in Treatment A because no enzyme had been added and highest in the warmed enzyme/substrate mixture of Treatment D. By the same rationale, Treatment C should have a lower rate than Treatment B because it is cooler. The predicted order is thus $A < C < B < D$. A suitable significance test would be a specific form of a nonparametric one-way analysis of variance using z as the test statistic.

11. The consultant could try a nonparametric two-way analysis of variance. The two levels of grouping would be 'housing

condition' and 'breed' with three groups at the first level and four at the second. A dozen or so samples for each combination of housing and breed would be useful, though the same number of samples should be used in each case since general rather than specific predictions are being tested. One thing the consultant should be careful to do is distribute pigs from different families arbitrarily across housing conditions so that any effect of housing on growth rate is not confounded with family-specific growth rates (related pigs might grow at a similar rate because they inherited similar growth characteristics). The analysis would indicate any independent effect of housing and breed and any interaction between the two in influencing growth rate.

12. Although the figures look very similar, they are deceptive because their y axes are scaled differently. The drop in numbers in felled deciduous forest represents 49% of the number in unfelled forest. In coniferous forest it represents only 38%. In proportional terms, therefore, the impact of felling seems to be greater in deciduous forest. However, the fact that the analysis is based on only single counts means it should act only as an exploratory analysis leading to a properly replicated confirmatory analysis.

13. A negative value of r_s indicates a negative correlation, i.e. the value of y decreases as that of x increases. The sign of the coefficient is ignored when checking against threshold values of significance.

14. Some predictions can be derived as follows:

(a) Difference predictions

Observation: The distribution of individuals across species varied between different sites.
Hypothesis: *Differences in the degree of dominance of species within a community vary with the ability of species to compete with others for limited resources.*
Prediction: *Dominant species will be those whose individuals win in contests with individuals of other species over the resource they occupy.*
Observation: Individuals of some species are smaller in some streams than in others and some streams are more polluted.
Hypothesis: *Pollution results in reduced body size among some freshwater species.*
Prediction: *The body size of any given species will be smallest in the most polluted stream and largest in the least polluted stream.*

(b) A trend prediction

Observation: Fewer organisms were seen attached to the substrate in fast flowing parts of the streams.
Hypothesis: *Flow rate influences the ability of organisms to settle on the substrate.*
Prediction: *If clean substrate is provided in areas of different flow rate, then fewer of the organisms drifting by will settle the faster the flow.*

15. Differences.

16. Yes, a 1 × 4 chi-squared analysis could be carried out, but care would have to be taken in calculating expected values because of the different surface areas of the body sites and their possibly different degrees of vulnerability.

17. Since the experimenter had collected ten counts for each combination of 'parent' and 'sibling' treatment, a two-way analysis of variance would provide most information. It would allow either general or specific predictions about the effects of 'parent' and 'sibling' treatments and the interaction between them to be tested. The ten values in each case **could** be totalled and used in a 2 × 2 chi-squared analysis but this would only test for an overall combined effect of the treatments; much useful information would thus be lost in comparison with the analysis of variance.

18. The threshold probability of 0.05 is an arbitrarily agreed compromise between the risk of **accepting** a null hypothesis in error (as would happen by setting the threshold p-value too high) and the risk of **rejecting** it in error (by setting the threshold too low). For many situations in biology, a threshold of $p = 0.01$ would result in an unnecessary risk of accepting the null hypothesis when it was not true.

19. Clearly this is not a sensible procedure because it confounds the size of prey with the amount each barracuda has already eaten. Barracuda may give up at a certain size of fish simply because they are satiated, not because the fish is too big. Also size is confounded with species, some of which may be distasteful or be unpalatable in other ways. Again, therefore, barracuda may reject a fish for reasons other than size.

20. There are at least three logical flaws in this line of reasoning. First, if males are generally larger than females then their brains will be proportionally larger too; any comparison of

brain sizes should thus be on a relative scale. Second, should the brains of males turn out to be relatively larger, there is no *a priori* reason to suppose this will affect learning or any other ability. Third, even if differences in brain size do produce differences in learning, training may overcome any such differences.

INDEX